AF517838

ENCYCLOPAEDIA OF ENGINEERING

ENCYCLOPAEDIA OF ENGINEERING

VOL. 5

MECHANICAL ENGINEERING

By

A.M. Kharmawphlang
A. Khongphai

AKANSHA PUBLISHING HOUSE
NEW DELHI

AKANSHA PUBLISHING HOUSE
4649-B/21, Ansari Road
Darya Ganj, New Delhi-110 002
Email: ektabooks@yahoo.com
Ph.: 23263193/9811582579
Fax: 011-23263193

Encyclopaedia of Engineering

First Published 2006

ISBN 81-8370-048-9 (Set)

PRINTED IN INDIA

Published by M.P. Misra for Akansha Publishing House, New Delhi and Printed at Aryan Enterprise, Delhi.

PREFACE

Engineering is the application of science to the needs of humanity. This is accomplished through the application of knowledge, mathematics, and practical experience to the design of useful objects or processes.

Engineering involves the knowledge of mathematical and natural sciences (biological and physical) gained by study, experience, and practice, applied with judgement and creativity to develop ways to utilize the materials and forces of nature for the benefit of mankind.

Encyclopaedia of engineering contains engineering terms with clear and concise definitions. The encyclopaedia is highly illustrated and written in plain, non-technical language and extensive cross-referencing makes concepts more easily understood. Timings and Twigg focus on a wide range of engineering disciplines — engineering science, electrical and electronic engineering, workshop practices and mechanical engineering — and provide definitions that are relevant to all.

This vital reference offers a wealth of essential information in a convenient, quick-find format. Whether you're a professional, a student, a writer, or a general reader with an interest in science, there is no better or more authoritative way to stay up-to-speed with the current language of the various branches of engineering or gain an understanding of their key ideas and concepts.

Encyclopaedia of Engineering is written specifically for further education students by experienced authors who have taught and written extensively at this level.

A

Abandonment Costs. The costs associated with abandoning a well or production facility. Such costs are specified in the authority for expenditure (AFE), and typically cover the plugging of wells; removal of well equipment, production tanks and associated installations; and surface remediation.

Abbit Metal. See Bearing, Babbit.

ABMA. American Boiler Manufacturers Association.

Abnormal Events. A term to indicate features in seismic data other than reflections, including events such as diffractions, multiples, refractions and surface waves. Although the term suggests that such events are not common, they often occur in seismic data.

Abnormal Pressure. 1. A subsurface condition in which the pore pressure of a geologic formation exceeds or is less than the expected, or normal, formation pressure. When impermeable rocks such as shales are compacted rapidly, their pore fluids cannot always escape and must then support the total overlying rock column, leading to abnormally high formation pressures. Excess pressure, called overpressure or geopressure, can cause a well to blow out or become uncontrollable during drilling. Severe underpressure can cause the drillpipe to stick to the underpressured formation. 2. Reservoir pore fluid pressure that is not similar to normal saltwater gradient pressure. The term is usually associated with higher than normal pressure, increased complexity for the well designer and an increased risk of well control problems. Pressure gradients in excess of around 10 pounds per gallon equivalent fluid density (0.52 psi/foot of depth) are considered abnormal. Gradients below normal are commonly called subnormal.

Abrasion Resistance. The ability of a material to resist surface wear.

Abrasion Test. A laboratory test to evaluate drilling-grade weighting material for potential abrasiveness. The test measures weight loss of a specially shaped, stainless-steel mixer blade after 20 minutes at 11,000 rpm running in a laboratory-prepared mud sample. Abrasiveness is quantified by the rate of weight loss, reported in units of mg/min. Mineral hardness, particle size and shape are the main parameters that affect abrasiveness of weighting materials. Some crystalline forms of hematite grind to a higher percentage of large particles than do other forms and are therefore more abrasive. Hematites are harder than barites, grind courser and are more abrasive. Thus, a hematite that is proposed as a weighting material for mud is typically a candidate for abrasion testing.

Abrasion. A process where hard particles are forced against and moved along a solid surface.

Abrasion. The wearing away of a surface by rubbing, as with sandpaper on wood.

Abrasive Erosion. Erosive wear caused by the relative motion of solid particles which are entrained in a fluid, moving nearly parallel to a solid surface.

Abrasive Jetting. A wellbore treatment in which a fluid laden with solid particles is used to remove deposits from the surface of wellbore tubulars and completion components. The treatment fluid is pumped at high pressure through a downhole tool equipped with nozzles that direct a jet, or jets, of fluid onto the target area. Most tool designs use a controlled rotary motion to ensure complete circumferential treatment of internal surfaces. Abrasive jetting techniques can also be used to cut completion or wellbore components. For this application, highly abrasive particles, such as sand, are carried in a fluid and jetted at the target area over an extended period to erode the tubular.

Abrasive Wheel. A grinding wheel composed of an abrasive grit and bonding agent.

Abrasive. A hard and wear-resistant material (commonly a ceramic) that is used to wear, grind, or cut away other material.

Absolute Age. The measurement of age in years. The determination of the absolute age of rocks, minerals and fossils, in years before the present, is the basis for the field of geochronology. The measurement of the decay of radioactive isotopes, especially uranium, strontium, rubidium, argon and carbon, has allowed geologists to more precisely determine the age of rock formations. Tree rings and seasonal sedimentary deposits called varves can be counted to determine absolute age. Although the term implies otherwise, "absolute" ages typically have some amount of potential error and are inexact. Relative age, in contrast, is the determination of whether a given material is younger or older than other surrounding material on the basis of stratigraphic and structural relationships, such as superposition, or by interpretation of fossil content.

Absolute Filter. A type of high-specification fluid filter frequently used to remove small solid particles from workover or treatment fluids that may be injected into, or placed adjacent to, the reservoir formation. In using absolute filters, all particles larger than the micron rating of the filter element in use will be removed from the treated fluid.

Absolute Humidity. Amount of moisture in the air, indicated in kg/kg of dry air.

Absolute Open Flow Potential. The maximum flow rate a well could theoretically deliver with zero pressure at the middle of the perforations. The term is commonly abbreviated as AOFP or OFP.

Absolute Permeability. The measurement of the permeability, or ability to flow or transmit fluids through a rock, conducted when a single fluid, or phase, is present in the rock. The symbol most commonly used for permeability is k, which is measured in units of darcies or millidarcies.

Absolute Pressure. Air at standard conditions (70□F air at sea level with a barometric pressure of 29.92 in Hg) exerts a pressure of 14.696 psi. This is the pressure in a system when the pressure gauge reads zero. So the absolute pressure of a system is the gauge pressure in pounds per square inch added to the atmospheric pressure of 14.696 psi (use 14.7 psi in environmental system work) and the symbol is "psia"

Absolute Pressure. The measurement of pressure relative to the pressure in a vacuum, equal to the sum of the pressure shown on a pressure gauge and atmospheric pressure.

Absolute Pressure. Total pressure measured from an absolute vacuum. It equals the sum of the gauge pressure and the atmospheric pressure corresponding to the barometer.

Absolute Temperature Scale. A scale of temperature measurement in which zero degrees is absolute zero.

Absolute Volume. The volume a solid occupies or displaces when added to water divided by its weight, or the volume per unit mass. In the oil field, absolute volume is typically given in units of gallons per pound (gal/lbm) or cubic meters per kilogram (m^3/kg).

Absolute Zero Temperature. Temperature measured from absolute zero (-459.67□F, or -273.16□C).

Absolute Zero. A hypothetical temperature at which there is total absence of heat. Since heat is a result of energy caused by molecular motion, there is no motion of molecules with respect to each other at absolute zero. It is theoretically the coldest possible temperature.

Absorbent. A material which, due to an affinity for certain substances, extracts one or more such substances from a liquid or gaseous medium with which it contacts and which changes physically or chemically, or both, during the process. Calcium chloride is an example of a solid absorbent, while solutions of lithium chloride, lithium bromide, and ethylene glycols are liquid absorbents.

Absorber. That part of the low side of an absorption system, used for absorbing vapor refrigerant.

Absorbing Boundary Conditions. An algorithm used in numerical simulation along the boundary of a computational domain to absorb all energy incident upon that boundary and to suppress reflection artifacts.

Absorptance. The ratio of absorbed incident energy to the total energy to which a body is exposed.

Absorption Band. The range of wavelengths of energy that can be absorbed by a given substance.

Absorption Oil. A light liquid hydrocarbon used to absorb or remove the heavier liquid hydrocarbons from a wet gas stream. Absorption oil is also called wash oil.

Absorption Refrigeration System. One in which the refrigerant, as it is absorbed in another liquid, maintains the pressure difference needed for successful operation of the system.

Absorption Refrigerator. Refrigerator that creates low temperatures by using the cooling effect formed when a refrigerant is absorbed by chemical substance.

Absorption System. A refrigeration system in which the refrigerant gas evolved in the evaporator is taken up in an absorber and released in a generator upon the application of heat.

Absorption Tower. A tower or column, which effects contact between a rising gas and a falling liquid, so that part of the gas may be taken up by the liquid.

Absorption. 1. The conversion of one form of energy into another as the energy passes through a medium. For example, seismic waves are partially converted to heat as they pass through rock. 2. The property of some liquids or solids to soak up water or other fluids. The natural gas dehydration process uses glycols (liquids) that absorb the water vapor to finally obtain dehydrated gas. In the same way, light oil, also called "absorption oil," is used to remove the heavier liquid hydrocarbons from a wet gas stream to obtain dry gas.

Absorption. A process whereby a material extracts one or more substances present in an atmosphere or mixture of gases or liquids accompanied by the material's physical and/or chemical changes.

Abyss. Alternate form: abyssal

Abyssal. Pertaining to the depositional environment of the deepest area of the ocean basins, the abyss. The depositional energy is low, the abyssal plain is flat and nearly horizontal, and fine-grained sediments are deposited slowly by waning turbidity currents or from suspension in the water. The water is thousands of meters deep (> 2000 m) [6520 ft], so the water is cold and sunlight is minimal.

Acceleration Due to Gravity. The rate of increase in velocity of a body falling freely in a vacuum. Its value varies with latitude and elevation. The International Standard is 32.174 ft. per second per second.

Acceleration Period. In cavitation and liquid impingement erosion, the stage following the incubation period, during which the erosion rate increases from near zero to a maximum value.

Acceleration. The rate of change of velocity, as a function of time. Expressed in m/s.

Accelerator Source. Synonyms: neutron generator.

Accelerator. 1. A downhole tool used in conjunction with a jar that stores energy for rapid release when the jar is fired. Depending on the operating mode, the energy in tension or compression is stored by means of a mechanical spring or compressible fluid such as nitrogen gas. To operate effectively, a jar and an accelerator should be selected as a matched set. 2. A downhole tool used in conjunction with a jar to store energy that is suddenly released when the jar operates. The energy provides an impact force that operates associated downhole tools or, in a contingency role, to help release a tool string that has become stuck. Accelerators are selected on the basis of their compatibility with the jar to be used.

Accelerometer. A device used during surveying to measure the acceleration of a ship or aircraft, or to detect ground acceleration in boreholes or on the Earth's surface produced by acoustic vibrations.

Accelration. The time rate of change of velocity; i.e., the derivative of velocity; with respect to time.

Acceptable Weld. A weld that meets all of the requirements and the acceptance criteria prescribed by the welding specifications.

Accessible Hermetic. An assembly of motor and compressor, inside a single bolted housing unit.

Accommodation. Sequence stratigraphic term for the amount of space available for sediment accumulation. Dominant influences on the amount of accommodation, or accommodation space, include subsidence and eustasy.

Accretion. The mechanism by which partially hydrated cuttings stick to parts of the bottomhole assembly and accumulate as a compacted, layered deposit.

Accumulation. 1. The phase in the development of a petroleum system during which hydrocarbons migrate into and remain trapped in a reservoir. 2. An occurrence of trapped hydrocarbons, an oil field.

Accumulator. A device used in a hydraulic system to store energy or, in some applications, dampen pressure fluctuations. Energy is stored by compressing a precharged gas bladder with hydraulic fluid from the operating or charging system. Depending on the fluid volume and precharge pressure of the accumulator, a limited amount of hydraulic energy is then available independent of any other power source. Well pressure-control systems typically incorporate sufficient accumulator capacity to enable the blowout preventer to be operated with all other power shut down.

Accumulator. Storage tank which receives liquid refrigerant from evaporator and prevents it from flowing into suction line before vaporizing.

Accuracy. The closeness of the agreement between the result of the measurement and the conventional true value of the quantity. Accuracy should not be confused with precision. Core measurements have well-defined calibration techniques and standards. Logging measurements are characterized during tool design and construction, and calibrated regularly to some standard. The quoted accuracy of a log then depends on the initial characterization, the reproducibility of the standard, and the stability of the measurement between calibrations and under downhole conditions. The actual accuracy also depends on the equipment performing and being operated to specification.

Acetic Acid. An organic acid used in oil- and gas-well stimulation treatments. Less corrosive than the commonly used hydrochloric acid, acetic acid treatments can be more easily inhibited or retarded for treatments of long duration. This is necessary particularly in applications requiring the protection of exotic alloys or in high-temperature wells. In most cases, acetic acid is used in conjunction with hydrochloric acid and other acid additives.

Acetone. A filler added to acetylene cylinders, capable of absorbing 25 times its own volume of acetylene.

Acid Attack. Caused by an incomplete flushing after an acid cleaning process of boilers or similar equipment.

Acid Cleaning. The process of cleaning the interior surfaces of steam generating units by filling the unit with dilute acid accompanied by an inhibitor to prevent corrosion and by subsequently draining, washing, and neutralizing the acid by a further wash of alkaline water.

Acid Condition in System. Condition in which refrigerant or oil in a system, is mixed with vapor and fluids that are acidic in nature.

Acid Effect. The change in a pulsed neutron capture measurement produced by acidizing a carbonate formation. Acidizing tends to increase the porosity as well as leave chlorides in the formation, thereby increasing the capture cross section. Both of these results affect the formation thermal decay time and must be taken into account in the interpretation.

Acid Embrittlement. A form of hydrogen embrittlement that may be induced in some metals by an acid.

Acid Frac. A hydraulic fracturing treatment performed in carbonate formations to etch the open faces of induced fractures using a hydrochloric acid treatment. When the treatment is complete and the fracture closes, the etched surface provides a high-conductivity path from the reservoir to the wellbore.

Acid Gas. A gas that can form acidic solutions when mixed with water. The most common acid gases are hydrogen sulfide [H_2S] and carbon dioxide [CO_2] gases. Both gases cause corrosion; hydrogen sulfide is extremely poisonous. Hydrogen sulfide and carbon dioxide gases are obtained after a sweetening process applied to a sour gas.

Acid Gas. A gas that that forms an acid when mixed with water. In petroleum processing, the most common acid gases are hydrogen sulfide and carbon dioxide.

Acid Hob. See: acidize, matrix acidizing.

Acid Inhibitor. A chemical additive used to protect wellbore components and treatment equipment from the corrosive action of an acid. The type and concentration of acid inhibitors are determined by the type of metal to be protected and the specific wellbore conditions, such as temperature and the length of exposure time anticipated during the treatment. To ensure efficient protection, the inhibitor should be consistently blended throughout the treatment fluid.

Acid Rain. Atmospheric precipitation with an pH below 5.6 to 5.7.

Acid Soak. A method of acid cleaning, in which the acid is pumped into the boiler and rests there for a period of time.

Acid Stimulation. See: acidize, matrix acidizing.

Acid Tank. The rubber-lined vessel used to transport raw or concentrated acid to the wellsite. Some acid additives attack or degrade rubber. Consequently, acid treatment fluids are not generally mixed or transported in acid tanks, but are instead mixed in special batch tanks or continuously mixed as the treatment is pumped.

Acid Wash. A wellbore acid treatment designed to remove scale or similar deposits from perforations and well-completion components. Acid-wash treatments generally do not include injection of treatment fluid into the reservoir formation.

Acid. 1. Antonyms: alkaline. 2. A generic term used to describe a treatment fluid typically comprising hydrochloric acid and a blend of acid additives. Acid treatments are commonly designed to include a range of acid types or blends, such as acetic, formic, hydrochloric, hydrofluoric and fluroboric acids. Applications for the various acid types or blends are based on the reaction characteristics of the prepared treatment fluid.

Acid. A chemical substance that yields hydrogen ions when dissolved in water.

Acid. Literally hitter, but chemically the state of a water solution containing a high concentration of hydrogen ions.

Acidic. The reaction of a substance with water resulting in an increase in concentration of hydrogen ions in solution (see acid).

Acidified. The addition of an acid (usually nitric or sulfuric) to a sample to lower the pH below 2.0. The purpose of the acidification is to "fix" a sample so it will not change until it is analyzed.

Acidity. Antonyms: alkalinity.

Acidity. Represents the amount of free carbon dioxide, mineral acids, and salts (especially sulfates of iron and aluminum) which hydrolyze to give hydrogen ions in the water. The acidity is reported as millie equivalents per liter of acid, or ppm acidity as calcium carbonate, or pH, the measure of hydrogen ion concentration.

Acidize. See: matrix acidizing.

Acidizing. The pumping of acid into the wellbore to remove near-well formation damage and other damaging substances. This procedure commonly enhances production by increasing the effective well radius. When performed at pressures above the pressure required to fracture the formation, the procedure is often referred to as acid fracturing.

Acoustic Basement. The portion of the Earth below which strata cannot be imaged with seismic data, or the deepest relatively continuous reflector. Acoustic basement, in some regions, coincides with economic basement and geologic basement, or that portion of the Earth that does not comprise sedimentary rocks.

Acoustic Coupler. An obsolete piece of equipment that converts acoustic signals from analog to electrical form and back. A common use of an acoustic coupler was to provide an interface between a telephone and an early type of computer modem.

Acoustic Emission. A measure of integrity of a material determined by sound emission when a material is stressed.

Acoustic Emission. A type of elastic wave produced by deformation or brittle failure of material and characterized by relatively high frequency.

Acoustic Impedance Section. A seismic reflectivity section, or a 2D or 3D seismic line, that has been inverted for acoustic impedance. Sonic and density logs can be used to calibrate acoustic impedance sections.

Acoustic Impedance. The product of density and seismic velocity, which varies among different rock layers, commonly symbolized by Z. The difference in acoustic impedance between rock layers affects the reflection coefficient.

Acoustic Log. 1. A display of traveltime of acoustic waves versus depth in a well. The term is commonly used as a synonym for a sonic log. Some acoustic logs display velocity. 2. A record of some acoustic property of the formation or borehole. The term is sometimes used to refer specifically to the sonic log, in the sense of the formation compressional slowness. However, it may also refer to any other sonic measurement, for example shear, flexural and Stoneley slownesses or amplitudes, or to ultrasonic measurements such as the borehole televiewer and other pulse-echo devices, and even to noise logs.

Acoustic Mode. A situation in which acoustic energy that propagates in one direction is confined in the other two directions as, for example, a mode confined to an interface between two different materials or within the borehole. The Stoneley wave, tube wave and flexural mode have important applications in formation evaluation, while most of the others, such as the Rayleigh wave and the various guided borehole modes (normal, leaky and hybrid modes), are considered interference that must be filtered out. In y slow formations, leaky modes can help determine formation compressional slowness.

Acoustic Positioning. A method of calculating the position of marine seismic equipment. Range measurements are made whereby distance is equal to acoustic signal traveltime from transmitter to hydrophone multiplied by the speed of sound in water. When sufficient acoustic ranges with a proper geometric distribution are collected, location coordinates x, y and z of the marine seismic equipment can be computed by the method of trilateration (measuring the lengths of the sides of overlapping triangles). Acoustic positioning is commonly used in towed streamer and ocean-bottom cable seismic acquisition modes.

Acoustic Transducer. A device for transforming electrical energy into sound, or vice versa. In sonic logging applications, acoustic transducers are usually made of piezoelectric ceramic or magnetostrictive materials, and may be used as either receivers or transmitters in a frequency range between about 1 and 30 kHz. The transducers are excited as either monopoles, emitting or receiving sound in all directions, or dipoles, emitting or receiving in one plane. In ultrasonic logging applications, acoustic transducers are made of piezoelectric ceramic materials, and often are used in alternating transmitter/receiver (pulse-echo) mode, in a frequency range from a few hundred kilohertz to a few megahertz.

Acoustic Transparency. The quality of a medium whose acoustic impedance is constant throughout, such that it contains no seismic reflections. An example of an acoustically transparent medium is water.

Acoustic Traveltime. See: traveltime.

Acoustic Velocity Log. See: acoustic log.

Acoustic Velocity. See: velocity.

Acoustic Wave. General term for a P-wave.

Acoustic. 1. Pertaining to sound. Generally, acoustic describes sound or vibrational events, regardless of frequency. The term sonic is limited to frequencies and tools operated in the frequency range of 1 to 25 kilohertz. 2. In geophysics, acoustic refers specifically to P-waves in the absence of S-waves (i.e., in fluids, which do not support S-waves, or in cases in which S-waves in solids are ignored).

Acoustic. A term pertaining to sound, or the science of sound.

Acquisition Log. The log that is actually recorded while taking the measurements. It is distinct from a playback, which is produced later on from digital data.

Acquisition. The generation and recording of seismic data. Acquisition involves many different receiver configurations, including laying geophones or seismometers on the surface of the Earth or seafloor, towing hydrophones behind a marine seismic vessel, suspending hydrophones vertically in the sea or placing geophones in a wellbore (as in a vertical seismic profile) to record the seismic signal. A source, such as a vibrator unit, dynamite shot, or an air gun, generates acoustic or elastic vibrations that travel into the Earth, pass through strata with different seismic responses and filtering effects, and return to the surface to be recorded as seismic data. Optimal acquisition varies according to local conditions and involves employing the appropriate source (both type and intensity), optimal configuration of receivers, and orientation of receiver lines with respect to geological features. This ensures that the highest signal-to-noise ratio can be recorded, resolution is appropriate, and extraneous effects such as air waves, ground roll, multiples and diffractions can be minimized or distinguished, and removed through processing.

Across the Line. A method of motor starting, which connects the motor directly to the supply line on starting or running.

Acrylamide Polymer. A linear, nonionic polymer made of acrylamide monomers, $CH_2=CHCONH_2$. High molecular-weight polyacrylamides are used as selective flocculants in clear-water drilling, low-solids muds and wastewater cleanup. Polymers made of smaller molecules are used as clay deflocculants in water muds, which can contain hardness ions. Polyacrylamides are not nearly as sensitive to salinity and hardness as the anionic polyacrylates (SPA). Also, being nonionic, they are not as powerful for flocculation or deflocculation applications. Acrylamide polymers are, however, susceptible to hydrolysis and release ammonia under hot, alkaline conditions.

Acrylamide-acrylate Polymer. A linear copolymer of acrylate (anionic) and acrylamide (nonionic) monomers, also called partially-hydrolyzed polyacrylamide (PHPA). The ratio of acrylic acid to acrylamide groups on the polymer chain can be varied in manufacturing, as can molecular weight. Another variable is the base used to neutralize the acrylic acid groups, usually NaOH or KOH, or sometimes NH_4OH. A concentration of approximately 10 to 30% acrylate groups provides optimal anionic characteristics for most drilling applications. High-molecular weight PHPA is used as a shale-stabilizing polymer in PHPA mud systems. It is also used as clay extender, either dry-mixed into clay or added at the rig to a low-bentonite mud. PHPA can also be used to flocculate colloidal solids during clear-water drilling and for wastewater cleanup. Low molecular-weight PHPA is a clay deflocculant.

Acrylamido-methyl-propane Sulfonate Polymer. A copolymer of 2-acrylamido-2methyl propane sulfonate and acrylamide. AMPS polymers are highly water-soluble anionic additives designed for high-salinity and high-temperature water-mud applications. (Alkyl-substituted acrylamide can be used instead of ordinary acrylamide, which lessens its vulnerability to hydrolysis at high temperature and high pH.) Polymers from 0.75 to 1.5 MM molecular weight are suggested for fluid-loss control in these difficult muds.

Acrylate Polymer. Linear, anionic polymer made from the monomer acrylic acid, $CH_2{=}CHCOO^- H^+$. The acrylic acid groups are evenly spaced along the chain. Acrylic acid polymer neutralized with NaOH is sodium polyacrylate (SPA). Polyacrylates are best utilized in soft water with low salinity to achieve the best dispersion and full chain elongation. Even low concentrations of hardness ions, for example, Ca^{+2}, precipitate polyacrylates. Low molecular-weight polyacrylates are used as clay deflocculants. High molecular weight polymers are used for fluid-loss control and as a clay extender. As an extender, SPA is added to bentonite at the grinding plant. It is also used at the rig in low-solids muds. Divalent cations can negate its benefits as a clay extender. SPA is highly efficient when used to flocculate colloids in native-solids muds, clear-water muds and wastewater cleanup. The polymer chain links together colloidal solids that can be removed by gravity settling in shallow pits or by applying hydrocyclone, centrifuge or filtration techniques.

Acrylic. Synthetic resin made from acrylic acid or a derivative thereof; acrylics possess the property of transparency and offer flame resistance.

Action. Refers to the action of a controller. It defines what is done to regulate the final control element to effect control.

Activated Alumina. Chemical which is a form of aluminum oxide. It is used as a drier or desiccant.

Activated Carbon. Is a specially processed carbon, used as a filter drier. Commonly used to clean air.

Activated Rosin Flux. Mixture of rosin and small amounts of organic-halide activators or organic-acid activators.

Activated Sludge. An aerobic biological process for conversion of soluble organic matter to solid biomass, removable by gravity or filtration.

Activation Energy. The energy required to initiate a reaction, such as diffusion.

Activation Log. A log of elemental concentrations derived from the characteristic energy levels of gamma rays emitted by a nucleus that has been activated by neutron bombardment. The carbon-oxygen log, elemental capture spectroscopy log, pulsed neutron spectroscopy log, aluminum activation log and oxygen activation log are all examples of activation logs. However, the term is most commonly used to refer to the aluminum and oxygen activation logs, the latter also being known as a water-flow log.

Activator. Substance that enhances the ability of a flux to remove oxides and other contaminants from surfaces being joined.

Active Margin. A boundary of colliding lithospheric plates. The present subduction zones of the Pacific Rim, the older mountains of the Alps, and the Himalayas represent active margins.

Active Storage Pile. A method of stockpiling coal, sometimes called live storage. The pile is located outside the plant but adjacent to it, and usually contains four or five days of operating supply. The pile is not compacted, as it is not stored long enough to be exposed to the hazard of spontaneous combustion.

Active Sulfide. See: corrosion coupon, neutralization, sulfide.

Activity of Aqueous Solutions. The escaping tendency, or vapor pressure, of water molecules in an aqueous solution compared with that of pure water, typically abbreviated a_w. Activity is expressed mathematically as the ratio of two vapor pressures: $a_w = p/p_o$, where p is vapor pressure of the solution and p_o is vapor pressure of pure water. The ratio ranges from near 0 to 1.0 and corresponds to percent relative humidity (% RH) of air in equilibrium with the aqueous solution. For pure water, $a_w = p_o/p_o = 1.00$ and RH = 100%. By increasing the concentration of salt (or other solutes) in the solution, a_w decreases, because vapor pressure of the solution decreases. However, a_w never reaches zero. Known-activity, saturated-salt solutions are used to calibrate RH meters. Measuring RH of air above an oil mud is a simple way to measure the activity (salinity) of its water phase. Adjusting the salinity of the water phase is a way to control movement of water into or out of shales that are being drilled with an oil mud. Chenevert related a_w in oil mud to RH above the mud sample and devised a practical test using an electrohygrometer to measure RH, called the "Chenevert Method."

Actuator. The portion of a regulating valve, which converts mechanical, fluid, thermal, or electrical energy; into mechanical motion to open or close the valve seats or other such devices.

Acyclic Compound. Synonyms: aliphatic compound.

Additives. Chemicals added to fuel in very small quantities to improve and maintain fuel quality and/or to lower emissions.

Additivity. A property of semivariogram models. Any linear combination of admissible models with positive coefficients can be nested or added together. Generally, single models are used for modeling experimental semivariograms that are close in shape to one of the basic admissible models, or for the approximate fitting of complex structural functions. Nested models are used to better fit complex structural functions.

Adhesion. (1) The attractive force between adjacent surfaces in a frictional contact; (2) the state in which two surfaces are held together by interfacial forces.

Adiabatic Change. A change in the volume, pressure, or temperature of a gas, occurring without a gain of heat or loss of heat.

Adiabatic Compression. Compressing a gas without removing or adding heat.

Adiabatic Cooling. A method in which paramagnetic salts are pre-cooled, and then demagnetized, thereby producing further cooling.

Adiabatic Efficiency. The ratio of actual work output of a heat engine to the ideal output.

Adiabatic Expansion. The expansion of a gas, vapor, or liquid stream from a higher pressure to a lower pressure, with no change in enthalpy.

Adiabatic Process. A thermo-dynamic process in which no heat is extracted from or added to the system of the process.

Adiabatic Saturation Process. A process to determine absolute or relative humidity.

Adiabatic. Occurring with no addition or loss of heat from the system under consideration.

Adjacent Bed. Synonyms: shoulder bed, surrounding bed.

Adjustable Choke. A valve usually used in well control operations to reduce the pressure of a fluid from high pressure in the closed wellbore to atmospheric pressure. It may be adjusted (opened or closed) to closely control the pressure drop. Adjustable choke valves are constructed to resist wear while high-velocity, solids-laden fluids are flowing by the restricting or sealing elements.

Adjustable Differential. A means of changing the difference between the control cut-in and cutout points.

Adjustable Resistor. A resistor whose value can be mechanically changed, usually by the use of a sliding contact.

Adjusted Flow Time. The approximated flow time used for a well-test analysis when the flow rate varies before or during the test period. It is calculated as t = cumulative well production since the last extended shut-in period divided by the flow rate just before a well is shut in for a buildup test.

Adsorbent. A material which has the ability to cause molecules of gases, liquids or solids to adhere to its internal surfaces without changing the adsorbent physically or chemically. In water treatment, a synthetic resin possessing the ability to attract and to hold charged particles.

Adsorption Inhibitors. They are materials which caused them to be adsorbed on to the metal surface due to their polar properties.

Adsorption. The adhesion of the molecules of gases, dissolved substances, or liquids in more or less concentrated form, to the surface of solids or liquids with which they are in contact. Commercial adsorbent materials have enormous internal surfaces.

Adsorption. The property of some solids and liquids to attract a liquid or a gas to their surfaces. Some solids, such as activated charcoal or silica gel, are used as surfaces of adhesion to gather liquid hydrocarbons from a natural gas stream. To complete the process, the solids are treated with steam to recover the liquid hydrocarbons.

Advection. The transfer of heat by horizontal movement of air.

Advective Transport Modeling. A series of techniques that use geostatistical methods to determine fluid and contaminant flow in the subsurface. These techniques are used primarily to study contamination in ground water in environmental studies.

Aeolian. Alternate form: eolian.

Aeolotropy. See: anisotropy.

Aerated Layer. The surface or near-surface, unconsolidated sedimentary layer that has been subject to weathering and whose pores are air-filled instead of liquid-filled. An aerated layer typically has a low seismic velocity.

Aeration Cell. See Oxygen Cell.

Aeration. Exposing to the action of air, like blowing air through water before discharging to a river.

Aerobic. 1. Referring to a condition or a situation in which free oxygen exists in an environment. 2. Referring to a condition or a situation or a living creature, such as a bacteria, in which oxygen is required to sustain life.

Aerobic. A condition in which "free" or dissolved oxygen is present in water.

Aerodynamic Noise. Also called generated noise, self-generated noise; is noise of aerodynamic origin in a moving fluid arising from flow instabilities. In duct systems, aerodynamic noise is caused by airflow through elbows, dampers, branch wyes, pressure reduction devices, silencers and other duct components.

Aftercooling / Intercooling. Cooling the engine intake air after the turbocharger and prior to introduction into the cylinder. Aftercooling increases engine power and lowers NO_x emissions.

Afterflow. The flow associated with wellbore storage following a surface shut-in. When a well is first shut in at the surface, flow from the formation into the bottom of the wellbore continues unabated until compression of the fluids in the wellbore causes the downhole pressure to rise. If the wellbore fluid is highly compressible and the well rate is low, the afterflow period can be long. Conversely, high-rate wells producing little gas have negligible afterflow periods.

Aftertreatment Devices. Devices which remove pollutants from exhaust gases after the gas leaves combustion chamber (e.g., catalytic converters or diesel particulate filters). The term "exhaust gas aftertreatment" is considered derogatory by some in the emission control industry, but there is no consensus on the use of such alternatives as "post-combustion treatment" or "exhaust emission control".

AGC Time Constant. See: automatic gain control.

AGC. See: automatic gain control.

Agglomerate. The clustering together of a few or many particles into a larger solid mass.

Agglomeration. See: aggregation.

Aggradation. The accumulation of stratigraphic sequences by deposition that stacks beds atop each other, building upwards during periods of balance between sediment supply and accommodation. Alternate form: aggradational.

Aggradational. Alternate form: aggradation.

Aggregate. See: aggregation.

Aggregation. The formation of groups or clusters of particles (aggregates) in a fluid. In water or in water-base mud, clay particles form aggregates in a dehydrated, face-to-face configuration. This occurs after a massive influx of hardness ions into freshwater mud or during changeover to a lime mud or gyp mud. Aggregation results in drastic reductions in plastic viscosity, yield point and gel strength. It is part of wastewater cleanup and water clarification. Alum or polymers cause colloidal particles to aggregate, allowing easier separation.

Aging. A change in the properties of certain metals and alloys that occurs at ambient or moderately elevated temperatures after hot working, heat treatment, or a cold working operation.

Agitator. A device used to cause motion in confined fluids.

Air Blast Transformer. A transformer cooled by forcing a circulation of air around its windings.

Air Capacitor. Is a capacitor, which uses air as the dielectric between the plates.

Air Changes. A method of expressing the amount of air leakage into or out of a building or room in terms of the number of building volumes or room volumes exchanged.

Air Cleaner. A devise used to remove air borne impurities.

Air Coil. Coil on some types of heat pumps used either as an evaporator or a condenser.

Air Conditioner, Unitary. An evaporator, compressor, and condenser combination; designed in one or more assemblies, the separate parts designed to be assembled together.

Air Conditioner. They are basically refrigeration devices cooling air and rooms rather then food compartments.

Air Conditioning Unit. An assembly of equipment for the treatment of air so as to control, simultaneously, its temperature, humidity, cleanliness and distribution to meet the requirements of a conditioned space.

Air Conditioning, Comfort. The process of treating air so as to control simultaneously its temperature, humidity, cleanliness and distribution to meet the comfort requirements of the occupants of the conditioned space.

Air Conditioning. The process of treating air to simultaneously control its temperature, humidity, cleanliness, and distribution to meet the requirements of the conditioned space.

Air Cooler. A factory-encased assembly of elements whereby the temperature of air passing through the device is reduced.

Air Diffuser. A circular, square, or rectangular air distribution outlet, generally located in the ceiling and comprised of deflecting members discharging supply air in various directions and planes, and arranged to promote mixing of primary air with secondary room air.

Air Diffusion Aerators. They are aerators into which air is pumped into the water through perforated pipes, plates, or any other method.

Air Drill. See: air drilling.

Air Drilling. A drilling technique whereby gases (typically compressed air or nitrogen) are used to cool the drill bit and lift cuttings out of the wellbore, instead of the more conventional use of liquids. The advantages of air drilling are that it is usually much faster than drilling with liquids and it may eliminate lost circulation problems. The disadvantages are the inability to control the influx of formation fluids into the wellbore and the destabilization of the borehole wall in the absence of the wellbore pressure typically provided by liquids.

Air Duct. A tube or conduit for conveying air from one place to another.

Air Flotation. A process of accelerating sedimentation, by introducing air into the water, this lowers the density of the water, and increases the differences in the densities of the water and the suspended particles. (DAF), Dissolved Air Flotation.

Air Gap -The space between magnetic poles, or between the rotating and stationary assemblies in a motor or generator.

Air Gun. A source of seismic energy used in acquisition of marine seismic data. This gun releases highly compressed air into water. Air guns are also used in water-filled pits on land as an energy source during acquisition of vertical seismic profiles.

Air Handler. The fan blower, heat transfer coil, filter, and housing parts, of a system.

Air Infiltration. The leakage of air into a room through cracks in doors, windows, and other openings.

Air Purge. The removal of undesired matter by replacement with air.

Air Quality Management District (AQMD). Administrative districts organized in California to control air pollution. Nationwide in the U.S., AQMDs are parallel to the areas designated for classification against the National Ambient Air Quality Standards (NAAQS). Generally, AQMDs and their national parallel encompass multiple jurisdictions and closely follow the definition of Consolidated Metropolitan Statistical Areas and Metropolitan Statistical Areas.

Air Saturated. Moist air in which the partial pressure of the water vapor is equal to the vapor pressure of water at the existing temperature. This occurs when dry air and saturated water vapor coexist at the same dry-bulb temperature.

Air Sensing Thermostat. A thermostat unit, in which the sensing element is located in the refrigerated space.

Air Shooting. A method of seismic acquisition using charges detonated in the air or on poles above the ground as the source. Air shooting is also called the Poulter method after American geophysicist Thomas Poulter.

Air Standard. Air having a temperature of (20□C), a relative humidity of 36 percent, and under a pressure of 14.70 PSIA. The gas industry usually considers (16□C) as the temperature of standard air.

Air Toxics. Toxic air pollutants, as classified by pertinent regulations. Examples of substances classified as air toxics by the US Clean Air Act include acetaldehyde, benzene, 1,3-butadiene, formaldehyde, and polycyclic organic matter (POM). California air toxics regulations also classify diesel exhaust particulates as a toxic air contaminant.

Air Vent. Valve, either manual or automatic, to remove air from the highest point of a coil or piping assembly.

Air Washer. A water spray system or device for cleaning, humidifying, or dehumidifying the air.

Air Wave. A sound wave that travels through the air at approximately 330 m/s and can be generated and recorded during seismic surveying. Air waves are a type of coherent noise.

Air, Ambient. Generally the air surrounding the object.

Air, Dry. Air without contained water vapor.

Air, Outdoor. Air taken from outdoors and, therefore, not previously circulated through the system.

Air, Recirculated. Return air passed through the conditioner before being again supplied to the conditioned space.

Air, Reheating. In an air conditioning system, the final step in treatment, in the event the temperature is too low.

Air, Return. Air returned from conditioned or refrigerated space.

Air. A substance containing by volume approximately 78 - 79% nitrogen; 20.95% oxygen, .94% argon, traces of carbon dioxide, helium, etc.

Airborne Sound. Sound which reaches the point of interest by radiation through the air.

Air-Cooled Condenser. Heat of compression is transferred from condensing coils to surrounding air. This may be done either by convection or by a fan or blower.

Air-cut Mud. See: gas-cut mud.

Air-lock. An intermediate enclosed chamber of a vacuum or pressure system through which an object may be passed without materially changing the vacuum or pressure of the system.

Air-Sensing Thermostat. A thermostat unit, in which the sensing element is located in the refrigerated space.

Alclad. Composite wrought product comprised of an aluminum alloy core having one or both surfaces as metallurgically bonded aluminum or aluminum alloy coating that is anodic to the core, and thus electrochemically protects the core against corrosion.

Alcohol Brine. A water and alcohol solution, which remains as a liquid below 0^0C.

Alford Rotation. A processing technique to project formation shear data recorded in any two orthogonal directions into the fast and slow shear directions in the presence of shear-wave anisotropy. In the sonic logging application, a dipole transmitter excites a flexural mode that is recorded at one set of receivers that is in-line with the dipole and other receivers that are 90^0 out of line (the cross-dipole component). A similar recording is made of the wave from a second dipole transmitter, mounted orthogonally to the first. The flexural-wave velocity is closely related to the formation shear velocity, particularly at low frequencies and in hard formations. Using all four waveforms, the Alford rotation is used to determine the speed and direction of the fast and the slow shear wave.

Algae. A minute fresh water plant growth which forms a scum on the surfaces of recirculated water apparatus, interfering with fluid flow and heat transfer. Lower form of plant life, usually green and blue green algae appear in cooling water systems. They are responsible for copious amounts of stringy green slime masses.

Alias Filter. A filter, or a set of limits used to eliminate unwanted portions of the spectra of the seismic data, to remove frequencies that might cause aliasing during the process of sampling an analog signal during acquisition or when the sample rate of digital data is being decreased during seismic processing.

Aliasing. The distortion of frequency introduced by inadequately sampling a signal, which results in ambiguity between signal and noise. Aliasing can be avoided by sampling at least twice the highest frequency of the waveform or by filtering frequencies above the Nyquist frequency, the highest frequency that can be defined accurately by that sampling interval.

Alidade. A telescopic surveying device used to construct surface topographic and geologic maps in the field. The alidade is mounted on a plane table, which has a sheet of paper on which to draw the map, and an object or location is sighted through the alidade. The edge of the alidade is aligned in the azimuthal direction of the object or location. The vertical angle from which elevation of the location can be calculated is measured using the calibrated arc of the alidade.

Aliphatic Compound. One of a group of organic compounds of carbon (C) and hydrogen (H) in which the carbon atoms have linear, branched chain (open), or both types of structures. Aliphatics, as they are informally called, can be divided into paraffinic (saturated) and olefinic (unsaturated) chain types. The simplest aliphatic, paraffinic hydrocarbon is methane, CH_4. The simplest aliphatic, olefinic hydrocarbon is ethylene, C_2H_6. In drilling fluids, particularly oil-base muds, the amounts and types of hydrocarbon in the mud can be an important parameter in overall performance of the mud.

Alive. A term referring to a circuit in which a current is flowing. Also referred to as live.

Alkali Metal. A metal in group IA of the periodic table that form strongly alkaline hydroxides (lithium, sodium, potassium, rubidium, cesium, francium).

Alkali. A substance having marked basic properties. Applying to hydroxides of potassium, sodium, lithium, and ammonium. They turn red litmus to blue. Includes hydroxides of the alkaline earth metals of barium, strontium, and calcium.

Alkaline Boil-Out. Employed to remove oil and grease deposits from heating surfaces.

Alkaline Earth Metal. A metal in group IIA of the periodic table (beryllium, magnesium, calcium, strontium, barium, radium).

Alkaline Flooding. An enhanced oil recovery technique in which an alkaline chemical such as sodium hydroxide, sodium orthosilicate or sodium carbonate is injected during polymer flooding or waterflooding operations. The alkaline chemical reacts with certain types of oils, forming surfactants inside the reservoir. Eventually, the surfactants reduce the interfacial tension between oil and water and trigger an increase in oil production. Alkaline flooding is not recommended for carbonate reservoirs because of the abundance of calcium: the mixture between the alkaline chemical and the calcium ions can produce hydroxide precipitation that may damage the formation. Alkaline flooding is also known as caustic flooding.

Alkaline. Having a pH greater than 7.

Alkaline. Pertaining to an aqueous solution, such as a water-base mud, which has more hydroxyl ions (OH^-) than hydrogen ions (H^+) and pH greater than 7.

Alkalinity Test. A measure of the total amount of hydroxyl ions in a solution as determined by titration with standardized acid. This test is a well-known water-analysis procedure to estimate hydroxyl, carbonate and bicarbonate ion concentrations. There are two pH endpoints, P and M, in this titration, corresponding to phenolphthalein and methyl orange indicators. The "P" endpoint is at pH 8.3 and the "M" endpoint is at pH 4.3. Each is reported in units of cm^3 acid/cm^3 sample. For water samples and very simple mud filtrates, P and M data indicate OH^-, HCO_3^- and CO_3^{-2} concentrations, but an alkalinity test is unreliable for analyzing complex mud filtrates. The API has established standards for conducting alkalinity tests.

Alkalinity. A chemical property of an aqueous system that implies that there are more hydroxyl ions (OH^-) in the system, or a potential to produce more hydroxyl ions, than there are hydrogen ions (H^+), or potential to produce hydrogen ions.

Alkalinity. An expression of the total basic anions (hydroxyl groups) present in a solution. It also represents, particularly in water analysis, the bi-carbonate, carbonate, and occasionally, the borate, silicate, and phosphate salts which will react with water to produce the hydroxyl groups.

Allen Type Screw. A screw with a recessed hex shaped head.

Allochthon. Alternate form: allochthonous.

Allochthonous. Pertaining to materials, particularly rock masses, that formed somewhere other than their present location, and were transported by fault movements, large-scale gravity sliding, or similar processes. Autochthonous material, in contrast, formed in its present location. Landslides can result in large masses of allochthonous rock, which typically can be distinguished from autochthonous rocks on the basis of their difference in composition. Faults and folds can also separate allochthons from autochthons.

Allogenic. Pertaining to minerals or rock fragments that formed in one location but were transported to another location and deposited. Clastic sediments in a rock such as sandstone are allogenic, or formed elsewhere.

Allotriomorphic Crystal. A crystal having a normal lattice structure, but with an imperfect outward shape due to the influence of its surroudings.

Allotropy. The possibility of existence of two or more different crystal structures for a substance (generally an elemental solid).

Alloy Steel. Steel containing specific quantities of alloying elements (other than carbon)and commonly accepted amounts of manganese, copper, silicon, sulfur, and phosphorus).

Alloy. 1. A metallic substance that is composed of two or more elements. 2. A substance having metal properties and being composed of two or more chemical elements of which at least one is a metal.

Alluvial. Pertaining to the subaerial (as opposed to submarine) environment, action and products of a stream or river on its floodplain, usually consisting of detrital clastic sediments, and distinct from subaqueous deposition such as in lakes or oceans and lower energy fluvial deposition. Sediments deposited in an alluvial environment can be subject to high depositional energy, such as fast-moving flood waters, and may be poorly sorted or chaotic.

Alluvium. Material deposited in an alluvial environment, typically detrital sediments that are poorly sorted.

Alpha Iron. The body-centered cubic (BCC) form of pure iron, stable below 910 C (1670 F).

Alpha Processing. A technique for combining a measurement that has a high accuracy but low precision with another measurement of the same quantity that has a high precision but low accuracy in order to produce a result that is better than either alone. Alpha processing is used to improve the vertical resolution of neutron porosity and other dual-detector nuclear logs. The detector near the source has better precision than the far detector in the sense that it responds more precisely to vertical changes. However, the near detector is less accurate because it is more affected by the borehole environment. Alpha processing mathematically superimposes the rapid changes of the near detector on the slowly changing but accurate far detector to produce an accurate log with high vertical resolution.

The technique is also used to improve results from the carbon-oxygen log and other pulsed neutron spectroscopy measurements. Two methods are used to determine the carbon/oxygen ratio. The windows method counts the number of gamma rays within energy windows placed at the main peaks for carbon and oxygen. This method has good statistical precision but poor accuracy, as gamma rays from other elements contaminate these windows. The other method, spectral stripping, compares the total spectrum against standards for many elements, inverting the spectrum to obtain the yield for each element. This method is more accurate but has less statistical precision. Averaging over a number of measurements, alpha processing adjusts the windows result with the more accurate spectral stripping in order to obtain a precise and accurate result.

Altered Zone. A near-wellbore formation zone, a few inches thick, whose acoustic velocity has been affected by impregnation with drilling fluids, stress relief, or both. The acoustic velocity of the rock in the immediate vicinity of the borehole wall can be much slower than that in the virgin formation. To measure the formation velocity, it may be necessary to use a sonic logging tool that has a greater spacing between transmitter and receiver array (about 10 to 15 ft [3 to 4.5 m]) than the standard sonic tool (about 3 to 5 ft [0.9 to 1.5 m]). The altered zone may also give rise to different acoustic modes, for example the hybrid mode or a second Stoneley wave.

Alternating Current (AC). Electric current in which the direction of the current alternates or reverses. In a 60 Hertz (cycle) current, the direction of current flow reverses in 1/120th of a second; most commonly used current.

Alternative Fuel. Fuel other than petroleum diesel or gasoline.

Alternator. A device which converts mechanical energy, into alternating current.

Altimeter. An instrument used to measure the height above a reference point, such as ground or sea level.

Alum. 1. A series of double salts of aluminum sulfate and potassium sulfate. Alum is used as a colloidal flocculant in wastewater cleanup. 2. Is an aluminum sulfate or filter alum. Acts in the range of 5.0 to 8.0 pH.

Alumina. Aluminum oxide occasionally found as an impurity in water in very small amounts.

Aluminum Activation Log. A wireline log of the concentration by weight of aluminum in the formation, based on the principle of neutron activation. Aluminum (^{27}Al) can be activated by capturing relatively low-energy neutrons from a chemical source to produce the isotope ^{28}Al, which decays with a half-life of 2.3 minutes and emits a relatively easily detected 1.78 MeV gamma ray. A natural gamma ray spectrometer will detect this gamma ray along with the other natural gamma rays. If the natural gamma spectrum has been measured before activation, it can be subtracted from the spectrum after activation to give an estimate of Al content. Al is a relatively direct indicator of the volume of clay, since clay minerals are alumino-silicates.

Aluminum Stearate. The salt of aluminum hydroxide and stearic acid (saturated C-18 fatty acid) with the formula $Al(O_2C_{18}H_{35})_3$. It is a grease-like solid. When mixed with oil (for example, diesel oil) and the mixture sprayed onto the surface of a foamy water mud, it helps the gas bubbles break out of the mud.

Ambient Air Temperature. Temperature of fluid (usually air) which surrounds object on all sides.

Ambient Conditions. The conditions of temperature, pressure, and humidity, existing around an instrument.

Ambient Noise. The normal sound in a room or other location.

Ambient Temperature. The temperature at a point or area expressed as an average of the surrounding areas or materials. Ambient surface temperature is generally given to be 70 to 80°F [21 to 27°C]-an average of daily and seasonal variations.

American Society for Testing and Materials (ASTM). A non-profit organization that establishes specifications and standard test methods for a broad range of materials and products. ASTM standards are recognized as definitive guidelines for quality of motor fuels.

American Standard Pipe Thread. A type of screw or thread, commonly used on pipe fittings to assure a tight seal.

American Wire Gauge (AWG). A system used in the United States for measuring the size of solid wires.

Amides. A group of organic chemicals with the general formula RCO^-NH_2 formed from reactions of ammonia (NH_3) and a carboxylic acid, $RCOO^-H^+$. "R" groups range from hydrogen to various linear and ring structures. Amides and polyamides are emulsifiers and surfactants, many of which are made from fatty acids.

Amine. A chemical use in water treatment as a filming or neutralizing agent to protect the metal parts.

Amines. A group of organic chemicals that are analogs of ammonia (NH_3), in which either one, two or three hydrogen atoms of ammonia are replaced by organic radicals. General formulas are: (1) primary amines, RNH_2, (2) secondary amines, R_1R_2NH, (3) tertiary amines, $R_1R_2R_3N$ and quaternary amines, $R_1R_2R_3$ R_4N^+X (where X represents an anion). Amines are organic bases (mildly alkaline) and react with acids to form nitrogenous, organic salts. Amines made from fatty acids are emulsifiers and oil-wetting agents for oilfield chemicals.

Ammeter Shunt. A low-resistance conductor, placed in parallel (shunt) with an ammeter movement, so that most of the current flows through the shunt, and only a small portion flows through the ammeter. This extends the useful range of the meter.

Ammeter. An instrument for measuring the magnitude of electric current flow.

Ammonia. Chemical combination of nitrogen and hydrogen (NH_3). Ammonia is a very efficient refrigerant and identified as R-117.

Amorphous. A solid not having a repetitive three-dimensional pattern of atoms.

Amorphous. Non-crystalline, without long-range order.

Amott Imbibition Test. Synonyms: imbibition test.

Ampacity. A wire's ability to carry current safely, without undue heating. The term formerly used to describe this characteristic was current capacity of the wire.

Amperage. An electron or current flow of one coulomb per second, past a given point in a circuit.

Ampere Hour. A term used for rating battery capacity. As stated, an ampere for an hour; ex: 5 amperes for 20 hours = 100 amp hour on a 20 hr rating.

Ampere Turns. A term used to measure magnetic force. It represents the product of amperes, times the number of turns of the coil, in an electromagnet.

Ampere. The unit used for measuring the quantity of an electrical current flow. One ampere represent a flow of one coulomb per second.

Amplification. The process of obtaining an output signal greater than the input signal, through auxiliary power controlled by the input signal. The proce of increasing the strength, current, power, or voltage the signal.

Amplitude Anomaly. An abrupt increase in seismic amplitude that can indicate the presence of hydrocarbons, although such anomalies can also result from processing problems, geometric or velocity focusing or changes in lithology. Amplitude anomalies that indicate the presence of hydrocarbons can result from sudden changes in acoustic impedance, such as when a gas sand underlies a shale, and in that case, the term is used synonymously with hydrocarbon indicator.

Amplitude Distortion. See: distortion.

Amplitude Variation with Offset. Variation in seismic reflection amplitude with change in distance between shotpoint and receiver that indicates differences in lithology and fluid content in rocks above and below the reflector. AVO analysis is a technique by which geophysicists attempt to determine thickness, porosity, density, velocity, lithology and fluid content of rocks. Successful AVO analysis requires special processing of seismic data and seismic modeling to determine rock properties with a known fluid content. With that knowledge, it is possible to model other types of fluid content.

A gas-filled sandstone might show increasing amplitude with offset, whereas a coal might show decreasing amplitude with offset. A limitation of AVO analysis using only P-energy is its failure to yield a unique solution, so AVO results are prone to misinterpretation. One common misinterpretation is the failure to distinguish a gas-filled reservoir from a reservoir having only partial gas saturation ("fizz water"). However, AVO analysis using source-generated or mode-converted shear wave energy allows differentiation of degrees of gas saturation. AVO analysis is more successful in young, poorly consolidated rocks, such as those in the Gulf of Mexico, than in older, well-cemented sediments.

Amplitude. 1. The difference between the maximum displacement of a wave and the point of no displacement, or the null point. The common symbol for amplitude is a. 2. The amount of displacement of a seismic wavelet measured from peak to trough. 3. The maximum instantaneous value of alternating current or voltage. It can be in either a positive or negative direction. The greatest distance through which an oscillating body moves from the mid point.

AMPS. See: acrylamido-methyl-propane sulfonate polymer, copolymer.

Anaerobic. 1. The condition of an environment in which free oxygen is lacking or absent. 2. A description of organisms that can survive in the absence of oxygen, particularly bacteria. 3. Pertaining to systems, reactions or life processes of species, such as bacteria, in which atmospheric oxygen is not present or not required for survival.

Anaerobic. A condition in which "free" or dissolved oxygen is not present in the water.

Analog Circuit. Electrical circuit that provides a continuous relationship between its input and output.

Analysis. The process of determining the composition of a substance, by chemical or physical methods.

Anemometer. An instrument for measuring the velocity of a fluid.

Angle of Approach. The acute angle at which a wavefront impinges upon an interface, such as a seismic wave impinging upon strata. Normal incidence is the case in which the angle of incidence is zero, the wavefront is parallel to the surface and its raypath is perpendicular, or Normal, to the interface. Snell's law describes the relationship between the angle of incidence and the angle of refraction of a wave.

Angle of Incidence. The acute angle at which a raypath impinges upon a line normal to an interface, such as a seismic wave impinging upon strata. Normal incidence is the case in which the angle of incidence is zero, the wavefront is parallel to the surface and its raypath is perpendicular, or normal, to the interface. Snell's law describes the relationship between the angle of incidence and the angle of refraction of a wave.

Angle of Lag or Lead. The phase angle difference, between two sinusoidal wave forms having the same frequency.

Angle Valve. A type of globe valve design, having pipe openings at right angles to each other. Usually one opening on the horizontal plane and one on the vertical plane.

Angstrom Units. A unit of wave length, equal in length to one ten billionth.

Angular Dispersion. The variation of seismic velocity in different directions.

Angular Unconformity. A surface that separates younger strata from eroded, dipping, older strata and represents a gap in the geologic record.

Anhydrite. 1. [$CaSO_4$] A member of the evaporite group of minerals and the soft rock comprising anhydrite formed by precipitation of calcium sulfate from evaporation of seawater. Anhydrite can also form through the dehydration of gypsum, another sulfate mineral found in evaporites. Anhydrite may occur as a cap rock above salt domes. 2. The anhydrous mineral form of calcium sulfate, $CaSO_4$. (Gypsum,

$CaSO_4$□$2H_2O$, is the hydrated form.) The presence of anhydrite or gypsum in rock will influence the type of mud selected for drilling the rock because when $CaSO_4$ dissolves in a water mud, Ca^{+2} and SO_4^{-2} ions are formed. Although both ions are detrimental to freshwater mud properties, Ca^{+2} is the more harmful of the two.

Anhydrous Calcium Sulfate. A dry chemical, made of calcium, sulfur, and oxygen.

Aniline Point Test. A test to evaluate base oils that are used in oil mud. The test indicates if an oil is likely to damage elastomers (rubber compounds) that come in contact with the oil. The aniline point is called the "aniline point temperature," which is the lowest temperature (□F or □C) at which equal volumes of aniline ($C_6H_5NH_2$) and the oil form a single phase. The aniline point (AP) correlates roughly with the amount and type of aromatic hydrocarbons in an oil sample. A low AP is indicative of higher aromatics, while a high AP is indicative of lower aromatics content. Diesel oil with AP below 120□F [49□C] is probably risky to use in oil-base mud. The API has developed test procedures that are the standard for the industry.

Anion Interchange. The displacement of one negatively charged particle by another on an anion-exchange material. This principle is used with water treatment.

Anion. 1. A negatively charged ion such as the chloride ion (Cl^-). 2. A negatively charged ion. Clay surfaces, groups on polymer chains, colloids and other materials have distinct, negatively charged areas or ions. Anionic characteristics affect performance of additives and contaminants in drilling fluids, especially water muds, in which clays and polymers are used extensively. 3. A negatively charged non-metallic ion.

Anionic. Antonyms: cationic.

Anisotropic Formation. A formation with directionally dependent properties. The most common directionally dependent properties are permeability and stress. Most formations have vertical to horizontal permeability anisotropy with vertical permeability being much less (often an order of magnitude less) than horizontal permeability. Bedding plane permeability anisotropy is common in the presence of natural fractures. Stress anisotropy is frequently greatest between overburden stress and horizontal stress in the bedding plane. Bedding plane stress contrasts are common in tectonically active regions. Permeability anisotropy can sometimes be related to stress anisotropy.

Anisotropic. Alternate form: anisotropy.

Anisotropy. 1. Material characteristic of exhibiting different values of a property in different crystallographic directions. 2. Predictable variation of a property of a material with the direction in which it is measured, which can occur at all scales. For a crystal of a mineral, variation in physical properties observed in different directions is anisotropy. In rocks, variation in seismic velocity measured parallel or perpendicular to bedding surfaces is a form of anisotropy. (Compare with homogeneity.)

Anneal. To soften by heating and allowing to cool slowly.

Annealing Twin. A twin formed in a crystal during recrystallization.

Annealing. 1. A generic term used to denote a heat treatment where the microstructure and properties of a material are altered; it frequently refers to a heat treatment where a previously cold-worked metal is softened by allowing it to recrystallize. 2. A process of heat treating metal, to get the desired properties of softness and ductility, (easy to form).

Annubar. A device that uses Pitot tubes to measure the gas flow rate within a pipeline. The gas volume is calculated from the difference between the flowing pressure and the static pressure of the gas.

Annular Blowout Preventer. A large valve used to control wellbore fluids. In this type of valve, the sealing element resembles a large rubber doughnut that is mechanically squeezed inward to seal on either pipe (drill collars, drillpipe, casing, or tubing) or the openhole. The ability to seal on a variety of pipe sizes is one advantage the annular blowout preventer has over ram-type blowout preventers. Most blowout preventer (BOP) stacks contain at least one annular BOP at the top of the BOP stack, and one or more ram-type preventers below. While not considered as reliable in sealing over the openhole as around tubulars, the elastomeric sealing doughnut is required by API specifications to seal adequately over the openhole as part of its certification process.

Annular BOP. See: annular blowout preventer.

Annular Flow. A multiphase-flow regime in which the lighter fluid flows in the center of the pipe, and the heavier fluid is contained in a thin film on the pipe wall. The lighter fluid may be a mist or an emulsion. Annular flow occurs at high velocities of the lighter fluid, and is observed in both vertical and horizontal wells. As the velocity increases, the film may disappear, leading to mist or emulsion flow. When the interface between the fluids is irregular, the term wavy annular flow may be used.

Annular Gas Flow. A flow of formation gas in the annulus between a casing string and the borehole wall. Annular gas flows occur when there is insufficient hydrostatic pressure to restrain the gas. They can occur in uncemented intervals and even in cemented sections if the cement bond is poor. After cementing, as the cement begins to harden, a gel-like structure forms that effectively supports the solid material in the cement slurry. However, during this initial gelling period, the cement has no appreciable strength. Hence, with the solid (weighting) material now supported by the gel structure, the effective density of the slurry that the reservoir experiences falls rather suddenly to the density of the mix water of the cement, which is usually freshwater, whose density is 8.34 lbm/gal, or a gradient of 0.434 psi/ft of vertical column height. Various chemical additives have been developed to reduce annular gas flow.

Annular Pressure. Fluid pressure in the annulus between tubing and casing or between two strings of casing.

Annular Production. Production of formation fluids through the casing-tubing annulus.

Annular Space. The space surrounding one cylindrical object placed inside another, such as the space surrounding a tubular object placed in a wellbore.

Annular Velocity. 1. The speed at which drilling fluid or cement moves in the annulus. It is important to monitor annular velocity to ensure that the hole is being properly cleaned of cuttings, cavings and other debris while avoiding erosion of the borehole wall. The annular velocity is commonly expressed in units of feet per minute or, less commonly, meters per minute. The term is distinct from volumetric flow. 2. The linear velocity of a fluid passing through an annular space. The term critical annular velocity is often used to describe the flow rate or velocity at which entrained solids will be efficiently transported by the annular fluid. If the fluid velocity falls below the critical rate, there will be a risk of particles settling, forming beds or bridges that may obstruct the wellbore.

Annuli. Plural form of annulus.

Annulus. 1. The space between two concentric objects, such as between the wellbore and casing or between casing and tubing, where fluid can flow. Pipe may consist of drill collars, drillpipe, casing or tubing. 2. With reference to invasion, a region between the flushed zone and the undisturbed zone containing a buildup of formation water. The annulus forms during invasion and is caused by the different mobilities of oil and water. It only occurs in the presence of both, but is unstable and will dissipate vertically or horizontally with time. The annulus has approximately the same water saturation as the flushed zone but contains formation water. When the formation water is much more saline than the mud filtrate, the annulus forms a conductive ring around the borehole. This conductivity will cause an induction log to read too low a resistivity, by an amount that depends on its depth of investigation and the radius of the annulus. (Laterologs are little affected since they respond to resistivity, not conductivity.) Often a medium log will be more affected than a deep log so that an annulus can be detected by out-of-order curves?medium curves reading less than either shallow or deep. Array induction logs contain enough information to solve and correct for the effect of the annulus. 3. The space between two concentric pipe strings, such as between the production tubing and casing in a well. The term may also refer to the space between a pipe string and the borehole wall in an openhole completion or openhole drillstem test (DST).

Anode. 1. The half of a battery that is positively charged and to which anions migrate by electrostatic attraction. Half of an electrolytic corrosion cell in metal is called the "anode," from which metal dissolves, often leaving pits. The anode is the part of a corrosion cell in which oxidation occurs. 2. A protective device to prevent electrolytic corrosion. Anodes (often made of Mg or Al metal) are sacrificed intentionally to protect a steel system, such as a buried pipeline or offshore platform. 3. In electrolysis or electrochemical corrosion, a site where metal goes into solution as a cation leaving behind an equivalent of electrons to be transferred to the opposite electron, called the cathode. 4. The electrode of an electrolyte cell at which oxidation occurs; electrons flow away from the anode in the external circuit; it is usually at the electrode that corrosion occurs and metal ions enter solution.

Anodic Protection. A technique to reduce the corrosion rate of a metal by polarizing it into its passive region, where dissolution rates are low.

Anodic Reaction. Electrode reaction equivalent to a transfer of positive charge from the electronic to the ionic conductor; an anodic reaction is an oxidation process.

Anodizing. 1. Forming a conversion coating on a metal surface by anodic oxidation; most often applied to aluminum. 2. The treatment of a metal surface whereby the metal is made anodic.

Anomalous. Alternate form: anomaly.

Anomaly. 1. An entity or property that differs from what is typical or expected, or the measurement of the difference between observed and expected values of a physical property. Anomalies can be of great interest in hydrocarbon and mineral exploration because they often indicate hydrocarbon and mineral prospects and accumulations, such as geologic structures like folds and faults. Geochemical anomalies at the surface of the Earth can also indicate an accumulation of hydrocarbons at depth. Geophysical anomalies, such as amplitude anomalies in seismic data and magnetic anomalies in the Earth's crust, can also be associated with hydrocarbon accumulations. 2. The difference between a measured value and the value predicted by a theoretical model, such as the Bouguer anomaly.

Anoxic. Anaerobic. The condition of an environment in which free oxygen is lacking or absent.

ANSI. American National Standard Institute, which is an organization defining standards for computer language.

ANSI-B.31.1. Power Piping.

ANSI-B.31.5. Refrigeration Piping.

Anti Knock Value. A premature explosive combustion, as the detonation of the fuel air mixture in an internal combustion engine, produces a characteristic knock. The Anti Knock Value is the measure of its resistance to the condition which tends to produce this knock.

Antialias Filter. Synonyms: alias filter.

Anticipating Control. One which, by artificial means, is activated sooner than it would be without such means, to produce a smaller differential of the controlled property. Heat and cool anticipators are commonly used in thermostats.

Anticipators. A small heater element in two-position temperature controllers which deliberately cause false indications of temperature in the controller in an at tempt to minimize the override of the differential and smooth out the temperature variation in the controlled space.

Anticlinal Trap. A type of structural hydrocarbon trap whose closure is controlled by the presence of an anticline.

Anticlinal. Alternate Form: anticline.

Anticline. An arch-shaped fold in rock in which rock layers are upwardly convex. The oldest rock layers form the core of the fold, and outward from the core progressively younger rocks occur. Anticlines form many excellent hydrocarbon traps, particularly in folds with reservoir-quality rocks in their core and impermeable seals in the outer layers of the fold. A syncline is the opposite type of fold, having downwardly convex layers with young rocks in the core.

Anti-Corrosive Additive. A lubricant additive to reduce corrosion.

Antiferromagnetism. A phenomenon where complete magnetic moment cancellation occurs as a result of antiparallel coupling of adjacent atoms or ions; the macroscopic solid possesses no net magnetic moment.

Anti-Foam Additive (Foam Inhibitor). An additive used to reduce or prevent foaming.

Antifoam Agent. 1. A chemical additive used to prevent the formation of foam during the preparation of treatment fluids or slurries at surface. Excess foam created during the mixing process may cause handling and pumping difficulties and may interfere with the performance or quality control of the mixed fluid. Antifoam agents may also be used to break foams returned from the wellbore, following a treatment, in preparation for disposal of the fluids.

2. The reduction of carry over by the addition of polymerized esters, alcohol's, and amides. The antifoam agent is absorbed on the steam generating surface resulting in a hydrophobic condition, causing fewer but larger steam bubbles, which readily coalesce. These agents also weaken the wall of the bubble formed, causing them to quickly burst on the water surface.

Antifoam. See: defoamer.

Antifoulants. Are materials which prevent fouling from depositing on heat transfer equipment. Materials that prevent deposits forming; include anti- oxidants, metal coordinators, and corrosion inhibitors. Compounds that prevent deposition are surfactants. They act as detergents or dispersants.

Antifreeze. Compounds of glycol's or alcohol's, that lower the freezing point of cooling water systems.

Antioxidant. A substance which when added in small amounts to petroleum products, will delay or inhibit undesirable changes; such as the formation of gum, sludge, and acidity, which are brought about by oxidation.

Anti-Oxidant. An additive for the purpose of reducing the rate of oxidation and subsequent deterioration of the material.

Anti-Scuffing Lubricant. A lubricant that is formulated to avoid scuffing.

Anti-Seizure Property. The ability of the bearing material to resist seizure during momentary lubrication failure.

Antisqueeze. The effect on a laterolog whereby the current lines are no longer properly focused but spread out at a certain distance into the formation. The effect occurs opposite a high-resistivity bed with low-resistivity shoulders. The result is that laterolog devices, in particular deep devices, tend to read too low and have less depth of investigation. Shoulder-bed correction charts correct for these effects in certain well-defined situations, such as no invasion in horizontal beds with vertical wells.

Antithetic Fault. A minor, secondary fault, usually one of a set, whose sense of displacement is opposite to its associated major and synthetic faults. Antithetic-synthetic fault sets are typical in areas of normal faulting.

Anti-Wear Additive. A lubricant additive to reduce wear.

Antiwhirl Bit. A drill bit, usually polycrystalline diamond compact (PDC) type, designed such that the individual cutting elements on the bit create a net imbalance force. This imbalance force pushes the bit against the side of the borehole, which in turn creates a stable rotating condition that resists backwards whirling, wobbling and downhole vibration. Antiwhirl bits allow faster rates of penetration, yet achieve longer bit life than more conventional bits, which are not dynamically biased to run smoothly, are inherently unstable, are vibration-prone and thus have shorter lives. No bit is whirl-proof, however.

AOF. Abbreviation for absolute open flow.

AOFP. Abbreviation for absolute open flow potential.

Aperture. 1. A portion of a data set, such as seismic data, to which functions or filters are applied. Aperture time, for example, can be specified, such as a window from 1.2 to 2.8 seconds. 2. A mechanism to limit the affects of measurements on a device or system. In seismic data acquisition, the length of the spread has the effect of an aperture.

API Cement. One of several classes of cement manufactured to the specifications of the *American Petroleum Institute (API) Specification 10A*. Classes of API cement are A, B, C, D, E, F, G and H.

API Fluid-loss Test. Synonyms: low-pressure, low-temperature filtration test.

Api Gravity (Api Degree). A measure of density used in the US petroleum industry.

API Unit. The unit of radioactivity used for natural gamma ray logs. This unit is based on an artificially radioactive concrete block at the University of Houston, Texas, USA, that is defined to have a radioactivity of 200 American Petroleum Institute (API) units. This was chosen because it was considered to be twice the radioactivity of a typical shale. The formation is the primary standard for calibrating gamma ray logs. However, even when properly calibrated, different gamma ray tools will not necessarily have identical readings downhole because their detectors can have different spectral sensitivities. They will read the same only if the downhole formation contains the same proportions of thorium, potassium and uranium as the Houston standard. For example, logging-while-drilling (LWD) tools have thicker housings than wireline tools, causing a different spectral response to the three sources of radioactivity, and therefore a different total gamma ray reading in some formations. The nuclear well log calibration facility at the University of Houston, known as the API pits, was opened in 1959 for the calibration of natural gamma ray and neutron logs. A facility for calibrating natural gamma ray spectroscopy logs was added later.

API Water. The amount of mixing water specified in *API Specification 10A* for specification testing of cement to meet API requirements. This amount is not intended to be a guide for mix water requirements in field applications.

API. Abbreviation for American Petroleum Institute, a trade association founded in 1919 with offices in Washington, DC, USA. The API is sponsored by the oil and gas industry and is recognized worldwide. Among its long-term endeavors is the development of standardized testing procedures for drilling equipment, drilling fluids and cements, called API Recommended Practices ("RPs"). The API licenses the use of its monogram (logo), monitors supplier quality assurance methods and sets minimum standards for materials used in drilling and completion operations, called API Specifications ("Specs"). The API works in conjunction with the International Standards Organization (ISO).

Apparent (Emf). The apparent voltage, as measured by the drop in pressure due to a current flowing through a resistance.

Apparent Anisotropy. In seismic data, the ratio of the velocity determined from normal moveout (i.e., primarily a horizontal measurement) to velocity measured vertically in a vertical seismic profile or similar survey. Apparent anisotropy is of particular importance when migrating long-offset seismic data and analyzing AVO data accurately. The normal moveout velocity involves the horizontal component of the velocity field, which affects sources and receivers that are offset, but the horizontal velocity field is not involved in velocity calculations from vertically measured time-depth pairs.

Apparent Dip. The angle that a plane makes with the horizontal measured in any randomly oriented section rather than perpendicular to strike.

Apparent Matrix. A calculation of the properties of the solid fraction of a rock from the combination of two logs. For example, by combining the density and neutron porosity measurements, it is possible to compute an apparent matrix density; by combining neutron porosity and sonic measurements, it is possible to compute an apparent matrix traveltime. The computations assume a particular fluid, usually fresh water, and particular response equations. The results are often displayed as quicklook logs for lithology identification. The word matrix is used here in the formation evaluation sense of the term rather than the geological one.

Apparent Velocity. In geophysics, the speed of a wavefront in a certain direction, typically measured along a line of receivers and symbolized by v_a. Apparent velocity and velocity are related by the cosine of the angle at which the wavefront approaches the receivers:

Apparent Viscosity. 1. The viscosity of a fluid measured at a given shear rate at a fixed temperature. In order for a viscosity measurement to be meaningful, the shear rate must be stated or defined. 2. The viscosity of a fluid measured at the shear rate specified by API. In the Bingham plastic rheology model, apparent viscosity (AV) is one-half of the dial reading at 600 rpm (1022 sec^{-1} shear rate) using a direct-indicating, rotational viscometer. For example, a 600-rpm dial reading is 50 and the AV is 50/2, or 25 cp.

Apparent Watts. The product of volts times amperes, in an alternating current circuit.

Apparent Wavelength. The wavelength measured by receivers when a wave approaches at an angle. The relationship between true and apparent wavelength can be shown mathematically as follows:

Applied-potential Method. Synonyms: equipotential method.

Appraisal. The phase of petroleum operations that immediately follows successful exploratory drilling. During appraisal, delineation wells might be drilled to determine the size of the oil or gas field and how to develop it most efficiently.

Aqueous. Watery. A substance containing water.

Aquifer. 1. A body of rock whose fluid saturation, porosity and permeability permit production of groundwater. 2. A water-bearing portion of a petroleum reservoir with a water drive. 3. Any water-bearing formation encountered while drilling. Drillers often are concerned about aquifers and are required to take special precautions in the design and execution of the well plan to protect fresh-water aquifers from contamination by wellbore fluids. Water in aquifers can flow into the wellbore, contaminate drilling fluids and cause well control problems.

Aquifer. A porous, subsurface geological structure carrying or holding water.

Arc Brazing. A brazing process in which the heat required is obtained from an electric arc.

Arc Furnace. An electric furnace, in which heat is produced by an arc between two electrodes.

Arc Welding – 1. A group of welding processes which produce coalescence of metals by heating them with an arc, with or without the application of pressure, and with or without the use of a filler metal. 2. A process where coalescence is obtained by heating with an electric arc.

Arc. A flash, caused by an electric current ionizing a gas or vapor.

Archie Equation. A particular relation proposed by G.E. Archie between the formation factor (F) and porosity (*phi*), in which $F = 1 / phi^m$, where the porosity exponent, m, is a constant for a particular formation or type of rock. In the original work, Archie proposed that *m* lay between 1.8 and 2.0 for consolidated sandstones, and close to 1.3 for loosely consolidated sandstones. *m* was named the cementation exponent shortly afterwards. This relation is also known as the Archie II equation.

Archie Rock. A rock whose petrophysical properties are well described by the Archie equation with constant values for the porosity exponent and the saturation exponent. Such rocks typically have very little clay, a regular pore structure and high-salinity water. The term often is used to describe a rock that is petrophysically simple.

Archimides Principal. States that an upward force acting on a body wholly or partly submerged in a fluid, is equal to the weight of the fluid displaced, and acts through the center of gravity of the fluid displaced, or the center of buoyancy.

Area Open to Flow. The calculated flow area provided by perforations across a specific zone of interest. The resulting value is used to calculate pressure drops and fluid-flow performance.

Arenaceous. Describing sandy-textured rock or sediment. Arenaceous does not necessarily imply silica-rich, but rather particles of sand size, 0.625 to 2 mm, according to the Udden-Wentworth scale.

Argillaceous. Describing rocks or sediments containing particles that are silt- or clay-sized, less than 0.625 mm in size. Most have a high clay-mineral content, and many contain a sufficient percentage of organic material to be considered a source rock for hydrocarbon.

Armature Air Gap. The air space between the stationary and rotating parts of a motor or generator, through which magnetic lines of force pass.

Armature Back Ampere Turns. The magnetic field produced by current flowing in the armature winding, that opposes and reduces the number of magnetic lines of force, produced by the field magnets of a motor or generator.

Armature Bar. Copper bars used in place of wire windings, in large armatures, generators, or motors.

Armature Circuit. The path that the current takes, in flowing through the windings from one brush to another.

Armature Coil. The loop or coil of copper wire, placed on the armature core, and forming part of the winding.

Armature Core -The laminated iron part of the armature, formed from thin sheets or disks of steel, on which the windings are placed.

Armature Current. The current flowing from the armature of a generator, to the armature of a motor. Not including the current taken by the shunt field.

Armature Demagnetization. The reduction in the effective magnetic lines of force, produced by the armature current.

Armature Reaction. The effect, that the magnetic field produced by the current flowing in the armature, has on the magnetic field produced by the field coils.

Armature Regulating Resistors. Are resistors, designed to regulate the speed or torque of a loaded motor, by placing a resistance in the armature or power circuit.

Armature Resistance. The resistance of the wire used in the windings of the armature, measured between the rings or brushes, or from positive to negative terminals.

Armature Slot. The groove or slot in the armature core, into which the coils or windings are placed.

Armature Tester. Any device used for locating faults or defects in the armature winding.

Armature Varnish. Is a liquid put on the field and armature windings, to improve the insulation of the cotton covering on the wires.

Armature Winding. All of the copper wire placed on the armature, and through which the current flows.

Armature. the rotating part of an electric motor or generator. The moving part of a relay or vibrator.

Armor. The metal strands on the outside of a wireline logging cable. Typical cables have two layers of metal strands, one wound clockwise and the other counterclockwise. The armor gives the cable its strength. It is used as the current return in some electrical measurements.

Aromatic Content Test. One of two quantitative analysis procedures for measuring aromatic content of base oils for use in oil mud as proscribed by the API. Results of the two aromatic content tests may differ because of the way the two gas-chromatography techniques separate and identify aromatics. The Institute of Petroleum (IP) instruments are more widely available than the American Society for Testing and Materials (ASTM) instruments, and the Institute of Petroleum method measures trace levels better than the more complex ASTM method.

Aromatic Hydrocarbon. A type of compound containing hydrogen and carbon atoms arranged in a symmetrical 6-carbon ring structure with single (C-C) and double (C=C) bonds alternating around the ring. Rings are single, multiple or fused and can have other chemical groups attached in place of hydrogen. Benzene, C_6H_6 is the simplest single-ring aromatic, napthalene, $C_{10}H_8$, the simplest fused-ring aromatic and toluene is the simplest aromatic, having an alkyl side chain, C_6H_5-CH_3. Xylene, a common oilfield chemical, has two methyl side chains, C_6H_4-$(CH_3)_2$. Aniline is the simplest aromatic amine, C_6H_5-NH_2 and is used in the aniline point test. Aromatic hydrocarbons in oils used to prepare oil-base mud can damage elastomers and increase the toxicity of the fluid. The aniline point test is used for screening oils to infer aromatic content.

Aromatics. A group of hydrocarbons of which benzene is the parent. They are called "aromatics" because many of their derivatives have sweet or aromatic odorous.

Array Induction. An induction tool or log that consists of several mutually balanced arrays whose signals are recorded separately and combined in software to produce the response desired. Typically, there is one transmitter and five to ten pairs of receivers and bucking coils that are balanced to remove direct coupling. The signals are combined in a wide variety of ways to produce the responses desired, as for example, deep-reading, high vertical resolution or some combination of both. There are trade-offs in any response. For example, a deep-reading log typically will not have high vertical resolution. If it does, it will be more sensitive to the invasion condition and cave effect.

Array Laterolog. An electrode device with multiple current electrodes configured in several different ways to produce several different responses. A typical array consists of a central electrode emitting survey current, with multiple guard electrodes above and below it. Current is sent between different guard electrodes to achieve greater or less focusing. The greater the focusing, the greater the depth of investigation. About five basic measurements are obtained in this way. This hardware focusing may be further improved by software focusing, in which the signals from the basic measurements are superimposed mathematically to ensure proper focusing in a wide range of conditions.

Array Propagation Resistivity. A resistivity recorded by a measurements-while-drilling propagation tool consisting of an array of transmitters and receivers whose signals are recorded separately and combined by software to produce the response desired. In a typical design, five transmitters emit a signal, and the phase shift and attenuation between two receivers are recorded. The phase shifts and attenuations are combined in different ways to produce borehole-compensated logs with different depths of investigation and radial resolution.

Array Sonic. A type of acoustic logging tool that uses a large number of receivers, typically 4 to 12. Modern acoustic logging tools are designed to measure not only the compressional wave but also the shear and other acoustic waves generated by the transmitter. The separation and identification of these waves are facilitated by the use of an array of receivers placed about 6 in. [15 cm] apart?close enough to avoid aliasing but far enough to sample a significant moveout in the wave. The waveforms at each receiver are recorded and processed by signal processing techniques, such as slowness-time coherence, to measure the velocities of the different waves.

Array. 1. Generally, a geometrical configuration of transducers (sources or receivers) used to generate or record a physical field, such as an acoustic or electromagnetic wavefield or the Earth's gravity field. 2. A geometrical arrangement of seismic sources (a source array, with each individual source being activated in some fixed sequence in time) or receivers (a geophone or hydrophone array) that is recorded by one channel. 3. An arrangement or configuration of electrodes or antennas used for resistivity, induced polarization (IP), or other types of electromagnetic surveying. Resistivity arrays typically consist of two current electrodes and two potential electrodes and are distinguished by the relative separations between the electrodes. Examples are the dipole-dipole, Schlumberger and Wenner arrays. 4. In computing, code written to access data in more than one dimension according to a name and subscripts that correspond to each dimension.

Arrival Time. The elapsed time between the release of seismic energy from a source and its arrival at the receiver.

Arrival. An event or appearance of seismic data as a reflection, refraction, diffraction or other similar feature, or the time at which seismic data appear. An event in a seismic section can represent a geologic interface.

Articulated Pistons. Two-piece pistons incorporating an entirely separate piston crown or dome with a separate skirt, and linking the two together with the piston pin. Many 1994 and later engines incorporate steel crown/aluminum skirt articulated pistons.

Artificial Intelligence. The study of ideas that enable computers to do the things that make people seem intelligent. The term is commonly abbreviated as A.I. Many computer programs written for use in the oil field utilize "rule based" approaches to provide expert systems. The rules are taken from an expert working in the field and are written in a way that attempts to reproduce the knowledge and approaches used by that expert to solve a range of real problems. Most such programs are limited to specific areas such as dipmeter interpretation, electrofacies determination, reservoir characterization, blowout prevention, drilling fluid selection, etc. Sometimes expert systems are written in computer languages that easily handle "rules" such as LISP, but once fully tested are usually translated to BASIC, C or FORTRAN to be compiled into efficient applications or programs.

Artificial Lift. Any system that adds energy to the fluid column in a wellbore with the objective of initiating and improving production from the well. Artificial-lift systems use a range of operating principles, including rod pumping, gas lift and electrical submersible pumps.

Artificial Magnet. Is a manufactured magnet, which is distinguished from a natural occurring magnet.

As-delivered BTU. The number of BTUs in a cubic foot of natural gas. The natural gas heat energy (BTU) will depend mainly on its water content at the delivered pressure and temperature conditions.

ASME Appendix Section I. Explains matter which is mandatory, unless specifically referred to in the rules of the code, including formulas.

ASME Boiler Code. The boiler code listing standards, specified by the American Society of Mechanical Engineers, for the construction of boilers.

ASME. American Society of Mechanical Engineers.

Aspect Ratio. The ratio of the length to the width, of a rectangular air grille or duct.

Asphalt. A solid or nearly solid form of bitumen that can melt upon heating and contains impurities such as nitrogen, oxygen and sulfur. Asphalt forms naturally when the light components or volatiles of petroleum have been removed or evaporated.

Asphaltenes. Organic materials consisting of aromatic and naphthenic ring compounds containing nitrogen, sulfur and oxygen molecules. The asphaltene fraction of crude is defined as the organic part of the oil that is not soluble in straight-chain solvents such as pentane or heptane. Asphaltenes exist as a colloidal suspension stabilized by resin molecules (aromatic ring systems) in the oil. The stability of asphaltic dispersions depends on the ratio of resin to asphaltene molecules. The determination of the quantity of resin is important in estimating the potential damage created by asphaltenes. Asphaltene precipitates as a result of pressure drop, shear (turbulent flow), acids, solution carbon dioxide [CO_2], injected condensate, mixing of incompatible crude oils or other conditions or materials that break the stability of the asphaltic dispersion. For example, in matrix acidizing, iron ions in solution favor the precipitation of asphaltene deposits.

Asphaltic Crude. Petroleum with a high content of naphthenic compounds, such as asphaltenes. Asphaltic crude is also known as naphthene-based crude oil when the paraffin wax content is low.

Asphaltic Mud Additive. A group of high-viscosity or solid hydrocarbons obtained from naturally occurring deposits or from the residue of petroleum refining, commonly used as additives for oil-base and water-base muds. Molten asphalt can be further processed by heating and passing air through the melt to oxidize and polymerize its components. Cooled, air-blown asphalt is glassy and can be ground. It has a high softening point and polar sites that offer emulsion-stabilizing qualities and affinity for clays and shales.

Aspirating Psychrometer. A device which draws sample of air through it to measure humidity.

Aspiration. Production of movement in a fluid by suction created by fluid velocity.

Assignment. The sale, transfer or conveyance of all or a fraction of ownership interest or rights owned in real estate or other such property. The term is commonly used in the oil and gas business to convey working interest, leases, royalty, overriding royalty interests and net profit interests.

Asthenosphere. The relatively plastic layer of the upper mantle of the Earth on which the tectonic plates of the lithosphere move. The asthenosphere is approximately 200 km [124 miles] thick and, owing to its depth below the Earth's surface, warm (~ 1400 °C) [2640 °F] but not molten. Here the mantle deforms by plastic flow in response to applied pressures above 100 MPa [14,500 psi]. This zone is considered coincidental, at least below oceanic crust, with the low-velocity zone of the upper mantle.

ASTM. American Society for Testing and Materials.

ASTM. American Society for Testing and Materials. Some API procedures for drilling fluids are similar to ASTM procedures.

Asynchronous Generator. An induction generator.

Asynchronous Motor. An induction motor, whose speed is not synchronous with the frequency of the supply line.

Asynchronous. Not having the same frequency. Out of step, or phase.

Atactic. A type of polymer chain configuration where side groups are randomly positioned on one side of the chain or the other.

Atmosphere. Is the mixture of gases and water vapor surrounding the earth.

Atmospheric Corrosion. Corrosion (oxidization) resulting from exposure of susceptible materials to oxygen and moisture. Atmospheric corrosion is generally associated with surface storage conditions, or with upper wellbore annuli that may not be fluid-filled.

Atmospheric Electricity. Is static electricity, produced between cloud fronts in the atmosphere.

Atmospheric Pressure. Pressure exerted by the weight of the atmosphere; standard atmospheric pressure is 101.325 kPa or 1.01325 bars or 14.696 psia or 29.921 inches of mercury at sea level.

Atom Percent (at%). Concentration specification on the basis of the number of moles (or atoms) of a particular element relative to the total number of moles (or atoms) of all elements within an alloy.

Atom. The smallest complete particle of an element, which can be obtained, yet which retains all physical and chemical properties of the element.

Atomic Number. The number of protons found in the nucleus of an atom, of an element.

Atomic Packing Factor (APF). The fraction of the volume of a unit cell that is occupied by "hard sphere" atoms or ions.

Atomic Weight. The weight of an elementary atom, in relation to the weight of an atom of hydrogen. A hydrogen atom being taken as 1.00g.

Atomize. Process of changing a liquid to minute particles or a fine spray.

Attapulgite. A needle-like clay mineral composed of magnesium-aluminum silicate. Major deposits occur naturally in Georgia, USA. Attapulgite and sepiolite have similar structures and both can be used in salt-water mud to provide low-shear rate viscosity for lifting cuttings out of the annulus and for barite suspension. Attapulgite and sepiolite are sometimes called "salt gel." Attapulgite has no capability to control the filtration properties of the mud. For use as an oil-mud additive, the clay is coated with quaternary amine, which makes it oil-dispersible and provides gel structure but does not improve the filter cake, unlike organophilic bentonite clay.

Attemperating Water Impurity. Refers to turbine deposits caused by impurities in the water used for attemperators. (Chemicals used with attemperating water should be of the volatile type).

Attemperator. An apparatus for reducing and controlling the temperature of a superheater vapor or a fluid.

Attenuate. The removal of undesirable features, such as multiple events, from seismic data.

Attenuation Resistivity. The ability of a formation to resist electrical conduction, as derived from the reduction in amplitude of the electromagnetic wave generated in a propagation resistivity measurement. At the frequencies used and within the range of measurement, the attenuation depends almost solely on the resistivity, so that the former can be transformed to the latter with a simple algorithm. The transform also depends on transmitter/receiver spacings and tool design. For a 2-MHz measurement, a typical measurement range is 0.2 to 50 ohm-m. Above 50 ohm-m, the dependence of attenuation on resistivity is too small to measure accurately.

Attenuation. 1. The loss of energy or amplitude of waves as they pass through media. Seismic waves lose energy through absorption, reflection and refraction at interfaces, mode conversion and spherical divergence, or spreading of the wave. 2. The reduction in amplitude of an electromagnetic wave passing through the formation, usually measured in decibels/meter, dB/m. The term is used in particular with reference to the propagation resistivity log and the electromagnetic propagation log. 3. The sound reduction process in which sound energy is absorbed or diminished in intensity as the result of energy conversion from sound to motion or heat.

Attitude. The orientation of a planar or linear feature in three-dimensional space. Planar features that are not horizontal, such as tilted strata, are described by their strike, or the azimuth of the intersection of the plane with a horizontal surface, and the dip, or the magnitude of its inclination from a horizontal reference. The trend and plunge of linear features, such as the axis of a fold, describe the azimuth of the line and its deviation from horizontal.

Attribute. A measurable property of seismic data, such as amplitude, dip, frequency, phase and polarity. Attributes can be measured at one instant in time or over a time window, and may be measured on a single trace, on a set of traces or on a surface interpreted from seismic data. Attribute analysis includes assessment of various reservoir parameters, including hydrocarbon indicators, by techniques such as amplitude variation with offset (AVO) analysis.

Attrition. The rubbing of one particle against another in a resin bed; frictional wear that will affect the site of resin particles.

Audio Measurement. A technique for recording sound at different positions in the borehole to generate a noise log. The measurement technique uses a microphone to record signals in the audible range approximately 20 to 20,000 Hz . In some circumstances, the frequency of the signal can be related to the source of noise and the flow regime, while the amplitude of the signal can be related to the flow rate. The useful signal lies approximately between 100 and 5000 Hz, with lower frequencies generally representing background and mechanical noise. The measurement may record the total signal over all frequencies, the signal at a single frequency, or consist of a set of measurements over different frequency ranges.

Aulacogen. In plate tectonics, a failed rift arm. At the junctions of tectonic plates, three intersecting lithospheric plates typically are separated by "arms." Arms might be areas of rifting, convergence or transform faults (similar to strike-slip faults). The arm along which the motion that spreads the plates apart ceases is termed the failed arm, or aulacogen. Spreading or rifting along the other arms of the triple junction can form new oceanic basins, whereas the aulacogen can become a sediment-filled graben.

Austenite. Face-centered cubic iron; also iron and steel alloys that have the FCC crystal structure.

Austenitizing. Forming austenite by heating a ferrous alloy above its upper critical temperature - to within the austenite phase region from the phase diagram.

Authigenic. Pertaining to minerals or materials that grow in place with a rock, rather than having been transported and deposited. These include quartz, chlorite and other pore-filling minerals or cements that grow during diagenesis. Evaporite minerals are authigenic, or formed in situ.

Authochthon. Alternate form: autochthonous.

Authority for Expenditure. A budgetary document, usually prepared by the operator, to list estimated expenses of drilling a well to a specified depth, casing point or geological objective, and then either completing or abandoning the well. Such expenses may include excavation and surface site preparation, the daily rental rate of a drilling rig, costs of fuel, drillpipe, bits, casing, cement and logging, and coring and testing of the well, among others. This estimate of expenses is provided to partners for approval prior to commencement of drilling or subsequent operations. Failure to approve an authority for expenditure (AFE) may result in delay or cancellation of the proposed drilling project or subsequent operation.

Autochthonous. Materials, especially rock masses, that formed in their present location and have not been transported. Fault surfaces can separate indigenous rocks from allochthonous rocks, although some allochthonous rocks are clearly delineated by their differing composition.

Autocorrelation. The comparison of a waveform to itself. Autocorrelation is useful in the identification of multiples or other regularly repeating signals, and in designing deconvolution filters to suppress them.

Automatic Control. The process of using the differences, between the actual value, and desired value, of any variable, to take corrective action, without human intervention.

Automatic Controller. A device that measures the value of a measured variable, and operates to correct or limit the deviation from a selected reference. Both measuring and control applications.

Automatic Defrost. System of removing ice and frost from evaporators automatically.

Automatic Expansion Valve (AEV). A type of metering device that senses low-side pressure and modulates in order to maintain low-side pressure constant.

Automatic Extraction Unit Turbine. Bleeds off part of the main steam flow at one, two, or three points. Valved partitions between selected stages control the extracted steam pressure at the desired level. When extracted steam flowing through the unit does not produce enough shaft power to meet the demand, more steam flows through the turbine to exhaust. Located between steam supply and process steam headers.

Automatic Frost Control. Control which automatically cycles refrigerating system to remove frost formation on evaporator.

Automatic Gain Control. A system to control the gain, or the increase in the amplitude of an electrical signal from the original input to the amplified output, automatically. AGC is commonly used in seismic processing to improve visibility of late-arriving events in which attenuation or wavefront divergence has caused amplitude decay.

Automatic Governing System. A system which correlates steam flow, pressure, shaft speed, and shaft output, for any one turbine unit.

Automatic Ice Cube Maker. Refrigerating mechanism designed to automatically produce ice cubes in quantity.

Automatic Reset (Integration). Is a type of control, in which the controller output, changes at a rate proportional to the deviation or error. The output will continue to change as long as any deviation or error exists.

Automatic. Self-acting. Operating by its own mechanism, when activated by some triggering signal.

Automation. The employment of devices, which automatically control one or more functions.

Autotrack. To use computer software to pick a particular reflection or attribute in seismic data automatically. Autotracking can speed interpretation of 3D seismic data,but must be checked for errors, especially in areas of faulting and stratigraphic changes.

Autotracking. Alternate form: autotrack.

Autotransformer. A transformer in which both primary and secondary coils, have turns in common. The step up or step down of voltage, is accomplished by taps in common windings.

Auxiliary Contacts. A set of contacts that perform a secondary function, usually in relation to the operation of a set of primary contacts.

AV. See: apparent viscosity.

Avagadro's Hypothesis. States that equal volumes of different gases, at the same temperature and pressure, will contain equal numbers of molecules.

Avagadro's Number (N). The number of elementary units such as atoms, formula units, molecules, or ions, that constitute one mole of the said particle.

Average Reservoir Pressure. 1. The pressure that would be obtained if all fluid motion ceases in a given volume of reservoir. It also is the pressure to which a well will ultimately rise if shut in for an infinite period. 2. A volumetric average of the pressure exerted by the fluids inside the reservoir at a specific depletion stage. Average reservoir pressure can be measured only when the well is shut in.

Average Velocity. In geophysics, the depth divided by the traveltime of a wave to that depth. Average velocity is commonly calculated by assuming a vertical path, parallel layers and straight raypaths, conditions that are quite idealized compared to those actually found in the Earth.

Average. See: arithmetic mean.

Averaging Element. A thermostat sensing element which will respond to the average duct temperature.

AVO. See: amplitude variation with offset.

Axial Fan. Consists of a propeller or disc type of wheel within a cylinder that discharges air parallel to the axis of the wheel.

Axial Flow Compressor. Uses rotor blades shaped like airfoils, to bite into the air, speed it up, and push it into the subsequent stationary blade passages. These passages are shaped to form diffusers, that slow up the incoming air, and make it pressurize itself by catching up with the air ahead of it.

Axial Loading. The force acting along the axis of an object. In wellbore tubulars, axial loading is typically expressed as tension or compression and may result from applied conditions such as set-down-weight, or be induced by operating conditions or variations such as changes in temperature that cause expansion or contraction of components.

Axial Surface. In folded rocks, the imaginary surface bisecting the limbs of the fold. The axial surface is called the axial plane when the fold is symmetrical and the lines defined by the points of maximum curvature of each folded layer, or hinge lines, are coplanar.

Azeotrope. Having constant maximum and minimum boiling points.

Azeotropic Mixture. Example of azeotropic mixture - refrigerant R-502 is mixture consisting of 48.8 percent refrigerant R-22 and 51.2% R-115. The refrigerants do not combine chemically, yet azeotropic mixture provides refrigerant characteristics desired.

Azimuth. 1. The angle between the vertical projection of a line of interest onto a horizontal surface and true north or magnetic north measured in a horizontal plane, typically measured clockwise from north. 2. The compass direction of a directional survey or of the wellbore as planned or measured by a directional survey. The azimuth is usually specified in degrees with respect to the geographic or magnetic north pole. 3. The angle that characterizes a direction or vector relative to a reference direction (usually True North) on a horizontal plane. The azimuth is usually quoted in degrees from 0 to 359.

Azimuthal Density. A type of logging-while-drilling density log in which the density is measured at different azimuths around the drill collar. The density measurement is focused, so that when the collar rotates, the measurement sees different azimuths around the borehole. An average density can be calculated by summing all the azimuthal data. Alternatively, the data can be summed over different segments, for example in four quadrants, to give an azimuthal density in four directions. When the hole is overgauge, certain quadrants will be firmly pressed against the borehole wall, while others may have a significant standoff and too high a delta rho. The ?good? quadrants can then be chosen for formation evaluation.

Azimuthal Laterolog. A type of electrode device that is able to measure resistivity in different directions around the sonde. In most laterologs, the electrodes are cylinders that average the resistivity azimuthally around the sonde. In azimuthal laterologs, the electrode is segmented radially in several portions, each of which responds to the resistivity in the direction it is facing.

Azimuthal Resolution. An angle that characterizes the ability of an azimuthal logging measurement to resolve changes in different directions around the tool; alternatively, the smallest angle for which a significant change can be detected.

Azimuthal. Pertaining to being focused in one direction. An azimuthal, or azimuthally focused, measurement has one or more directions perpendicular to the surface of a logging tool from which it receives most of its signal. Examples are the density, laterolog and microresistivity logs. A nonazimuthal, or azimuthally symmetric measurement is one which measures equally in all directions around the tool. Examples are the induction, propagation resistivity log and gamma ray.

B

Back Pressure. Pressure in low side of refrigerating system; also called suction pressure or low-side pressure.

Back Seating. Fluid opening/closing such as a gauge opening or to seal the joint where the valve stem goes through the valve body.

Back Work Ratio. Is the fraction of the gas turbine work used to drive the compressor.

Background Noise. Sound other than the wanted signal. In room acoustics, the irreducible noise level measured in the absence of any building occupants.

Backing Ring. Backing in a form of a ring, generally used in welding of piping.

Backwash. The counter-current flow of water through a resin bed (that is, in at the bottom of the exchange unit, out at the top) to clean and regenerate the bed after exhaustion (water treatment). Also, the process whereby a filtering mechanism is cleaned by reversing the flow through the filter.

Bacteria. Microscopic unicellular living organisms.

Baffle. Plate or vane used to direct or control movement of fluid or air within confined area.

Baghouse. A chamber containing bags for filtering solids out of gases.

Bainite. An austenitic transformation product found in some steels and cast irons; it forms at temperatures between those at which pearlite and martensite transformations occur; the microstructure consists of alpha-ferrite and a fine dispersion of cementite.

Ballast Gas. Are the nonflammable portion of the gas, such as carbon dioxide.

Band Gap Energy. For semiconductors and insulators, the energies that lie between the valence and conduction bands; for intrinsic materials electrons are forbidden to have energies within this range.

Barometer. Instrument for measuring atmospheric pressure. It may be calibrated in pounds per square inch, in inches of mercury in a column in millimeters or kPa.

Base Metal. 1. The metal present in the largest proportion in an alloy. (Copper is the base metal in brass). 2. The substrate metal that is coated or protected by a surface coating.

Base. An alkaline substance.

Basicity. The ability of a substance to boost the pH after neutralizing all the acid species.

Batch Operation. The utilization of ion-exchange resins to treat a solution in a container wherein the removal of ions is accomplished by agitation of the solution and subsequent decanting of the treated liquid.

Baudelot Cooler. Heat exchanger in which water flows by gravity over the outside of the tubes or plates.

Bauxite. An ore of aluminum consisting of moderately pure hydrated alumina - $Al_2O_3 \cdot 2H_2O$.

Bearing, Air. A bearing using air as a lubricant.

Bearing, Aligning. A bearing with an external spherical seat surface that provides a compensation for shaft or housing deflection or misalignment.

Bearing, Annular. Usually a rolling bearing of short cylindrical form supporting a shaft carrying a radial load.

Bearing, Anti-Friction. A bearing containing a solid lubricant.

Bearing, Axial Load. See Bearing, Thrust.

Bearing, Babbit. A bearing metal of non-ferrous material, containing several tin-based alloys, mainly copper, antimony, tin and lead.

Bearing, Ball. A rolling element bearing in which the rolling elements are spherical.

Bearing, Big End. A bearing at the larger (crankshaft) end of a connecting rod in an engine.

Bearing, Bimetal. A bearing consisting of two layers.

Bearing, Bottom End. See Bearing, Big End.

Bearing, Bush. A plain bearing in which the lining is closely fitted into the housing in the form of a bush, usually surfaced with a bearing alloy.

Bearing, Circular Step. A flat circular hydrostatic bearing with a central circular recess.

Bearing, Fixed Pad. An axial or radial load bearing equipped with fixed pads, the surface of which a are contoured to promote hydrodynamic lubrication.

Bearing, Floating Ring. A type of journal bearing that includes a thin ring between the journal and the bearing. The ring floats and rotates at a fraction of the journal rotational speed.

Bearing, Floating. A bearing designed or mounted to permit axial displacement between shaft and housing.

Bearing, Fluid. See Hydrostatic Bearing.

Bearing, Full Journal. A journal bearing that surrounds the journal by a full 360□.

Bearing, Gas. A journal or thrust bearing lubricated with gas.

Bearing, Half Journal. A bearing extending 180□ around a journal.

Bearing, Journal. A machine part in which a rotating shaft revolves or slides.

Bearing, Magnetic. A type of bearing in which the force that separates the relatively moving surfaces is produced be a magnetic field.

Bearing, Main. A bearing supporting the main power-transmitting shaft.

Bearing, Mitchell -See Tilting Pad Bearing.

Bearing, Needle. A bearing in which the relatively moving parts are separated by long thin rollers that have a length-to-diameter ratio exceeding 5.0.

Bearing, Noncontact. A bearing in which no solid contact occurs between relatively moving surfaces.

Bearing, Pedestal. A bearing that is supported on a column or pedestal rather than on the main body of the machine.

Bearing, Pivot. An axial load bearing, radial-load-type bearing which supports the end of a shaft or pivot.

Bearing, Porous. Made from porous material, such as compressed metal powders, the pores acting either as reservoirs for holding or passages for supplied lubricant.

Bearing, Roller. A bearing in which the relatively moving parts are separated by rollers.

Bearing, Rubbing. A bearing in which the relatively moving parts slide without deliberate lubrication.

Bearing, Self-Aligning. A roller-element bearing with one spherical raceway that automatically provides compensation for shaft or housing deflection or misalignment.

Bearing, Self-Lubricating. A bearing independent of external lubrication. These bearings may be sealed for life after packing with grease or may contain self-lubricating material.

Bearing, Sleeve. A cylindrical plain bearing used to provide radial location for a shat, which moves axially. Sleeve bearings consist of one or more layers of bearing alloys, bonded to a steel backing.

Bearing, Slide. A bearing used or positioning a slide or for axial alignment of a long rotating shaft.

Bearing, Step. A plane surface bearing that supports the lower end of a vertical shaft.

Bearing, Thrust. A bearing in which the load acts in the direction of the axis of rotation.

Bearing, Tilting Pad. A pad bearing in which the pads are free to take up a position at an angle to the opposing surface according to the hydrodynamic pressure distribution over its surface.

Bearing, Trunnion. A bearing used as a pivot to swivel or turn an assembly.

Bed Depth. The height of the resinous material in the column after the ion exchanger has been properly conditioned for effective operation.

Bed Expansion. The effect produced during backwashing when the resin particles become separated and rise in the column. The expansion of the bed due to the increase in the space between resin particles may be controlled by regulating backwash flow (typical with water treatment).

Bed. A mass of ion-exchange resin particles contained in a column.

Bellows Seal. A type of mechanical seal that utilizes a bellows for providing secondary sealing.

Bellows. Corrugated cylindrical container which moves as pressures change, or provides a seal during movement of parts.

Bending Moment. The algebraic sum of the couples or the moments of the external forces, or both, to the left or right of any section on a member subjected to bending by couples or transverse forces, or both.

Bernoulli's Theorem. In stream of liquid, the sum of elevation head, pressure head and velocity remains constant along any line of flow provided no work is done by or upon liquid in course of its flow, and decreases in proportion to energy lost in flow.

Bicacarbonate Alkalinity. The presence in a solution of hydroxyl (OH-) ions resulting from the hydrolysis of carbonates or bicarbonates. When these salts react with water, a strong base and a weak acid are produced, and the solution is alkaline.

Bicarbonate. An ion or salt of carbonic acid, containing hydrogen, carbon, and oxygen (HCO_3), such as sodium bicarbonate, $NaHCO_3$.

Bi-Fueled Vehicle. A vehicle with two separated fuel systems designed to run on either conventional fuel or an alternative fuel using only one fuel at a time.

Bimetal Strip. Temperature regulating or indicating device which works on principle that two dissimilar metals with unequal expansion rates, welded together, will bend as temperatures change.

Bimetallic Couple. A joint or union of two dissimilar metals.

Biocide. A chemical used to control the population of troublesome microbes.

Biodiesel. The mono alkyl esters of long chain fatty acids derived from renewable lipid feedstocks, such as vegetable oils and animal fats, for use in compression ignition (diesel) engines. Manufactured by transestrification of the organic feedstock by methanol.

Black Liquor. The liquid material remaining from pulpwood cooking in the soda or sulfate papermaking process.

Blade Clearance (Steam Turbine). Reaction turbine - the gap between blade and casing. Impulse turbine - the gap between the stationary and moving blades.

Blast Freezer. Low-temperature evaporator which uses a fan to force air rapidly over the evaporator surface.

Blast Furnace Gas. Is the waste product from furnaces used to smelt iron ores.

Bleeder Valve. A valve designed to slowly relief a liquid or gas form system.

Bleeding. Slowly reducing the pressure of liquid or gas from a system or cylinder by slightly opening a valve.

Bleedoff Rate. The rate at which water is continuously removed from a system.

Bleedoff. The continuous removal of water from a re-circulating water system.

Block Copolymer. A linear copolymer in which identical mer units are clustered in blocks along the molecular chain.

Blowdown. In connection with boilers or cooling towers, the process of discharging a significant portion of the aqueous solution in order to remove accumulated salts, deposits and other impurities.

Body-centered Cubic (BCC). Within the cubic unit cell atoms are located at corner and center cell positions.

Boiler Feed Water. The total water fed to a boiler producing steam. This water is the mixture of return steam condensate and makeup water.

Boiler Horsepower. The work required to evaporate 34.5 lb of water per hour into steam from and at 100□C.

Boiler Lay-Up. Storing and protecting the boiler when not in use.

Boiler. Closed container in which a liquid may be heated and vaporized.

Boiling Out. The boiling of high alkaline water in boiler pressure parts for the removal of oil, greases, prior to normal operation or after major repairs.

Boiling Point. The temperature at which the vapor pressure of a liquid equals the absolute external pressure at the liquid-vapor interface.

Boiling Temperature. Temperature at which a fluid changes from a liquid to a gas.

Boiling. See Vaporization.

Bonding Energy. The energy required to separate two atoms that are chemically bonded to each other.

Bore. Inside diameter of a cylinder.

Bourdon Tube. Thin-walled tube of elastic metal flattened and bent into circular shape, which tends to straighten as pressure inside is increased. Used in pressure gauges.

Boyles' Law. If the temperature on a gas is constant, the volume is inversely proportional to the pressure. By formula - $VP = V_1P_1$.

Brake Mean Effective Pressure (BMEP). The work accomplished during one engine cycle divided by the engine swept volume. It is essentially the engine torque normalized by the engine displacement. The word "brake" denotes the actual torque/power available at the engine flywheel as measured on a dynamometer. Thus, BMEP is a measure of the useful power output of the engine.

Brake Specific Fuel Consumption (BSFC). BSFC is the ratio of the engine fuel consumption to the engine power output (as measured at the flywheel). BSFC has units of grams of fuel per kilowatt-hour (g/kWh) or pounds mass of fuel per brake horsepower-hour (lb/bhp-hr). BSFC is a measure of engine efficiency.

Branch Circuit. Wiring between the last over current device and the branch circuit outlets.

Branched Polymer. A polymer having a molecular structure of secondary chains that extend from the primary main chains.

Brass. A copper-zinc alloy containing up to 40% zinc and some smaller amounts of other metals.

Brass. Copper (Cu) based alloy with zinc (Zn) as the main alloying element.

Brayton Cycle (also referred to as the Joule Cycle). A rotating machine in which compression and expansion take place. Gas turbine are such an example.

Braze. A weld produced by heating an assembly to suitable temperatures and by using a filler metal having a liquid us above 450□C. The filler metal is distributed between the closely fitted facing surface of the joint by capillary action.

Brazing, Block. A brazing process in which the heat required is obtained from heated blocks applied to the parts to be joined.

Brazing. A metal joining technique that uses a molten filler metal alloy having a melting temperature greater than about 425 degrees C (800 F).

Breakthrough. The first appearance in the solution flowing from an ion-exchange unit of unabsorbed ions similar to those which are depleting the activity of the resin bed. Breakthrough is an indication that regeneration of the resin is necessary.

Brine. Water saturated with a chemical such as salt.

Brinell Hardness Test. A test for determining the hardness of a material by forcing a hard steel or carbide ball of specified diameter into it under a specified load; the result is expressed as the Brinell hardness number.

British Thermal Unit (BTU). The Btu is defined as the heat required to raise the temperature of a pound of water from 59□ to 60□F.

Brittle Crack Propagation. A very sudden propagation of a crack with the absorption of no energy except that stored elastically in the body.

Brittle Fracture. Separation of a solid with little or no macroscopic plastic deformation; fracture occurs by rapid crack propagation with less expenditure of energy than for ductile fracture; brittle tensile fractures have a bright, granular appearance and exhibit little or no necking; typical fracture mode of a glass or ceramic.

Brittleness. The tendency of a material to fracture without first undergoing significant plastic deformation.

Brittleness. The tendency of a material to fracture without first undergoing significant plastic deformation.

Bronze. A copper-rich copper tin alloy with or without small proportions of other elements.

Bronze. Copper (Cu) based alloy with tin (Sn) as the main alloying element.

BTU. British Thermal Unit. Is the measure of heat energy.

Buffer. A substance used in solution, which accepts hydrogen ions or hydroxyl ions, added to the solution as acids or alkali's, minimizing a change in pH.

Bulb. The name given to the temperature-sensing device located in the fluid for which control or indication is provided. The bulb may be liquid-filled, gas filled, or gas-and-liquid filled. Changes in temperature produce pressure changes within the bulb which are transmitted to the controller.

Bulge. A local distortion or swelling outward caused by internal pressure on a tube wall or boiler shell due to overheating.

Burgers Vector (*b*). A vector that denotes the magnitude and direction of lattice distortion associated with a dislocation.

Bus Bar. A heavy, rigid metallic conductor which carries a large current and makes a common connection between several circuits. Bus bars are usually uninsulated and located where the electrical service enters a building; that is, in the main distribution cabinet.

Bypass. 1. A pipe or duct, usually controlled by valve or damper, for conveying a fluid around an element of a system. 2. Passage at one side of, or around, a regular passage.

Bypass Feeder. A closed tank that is installed in a system in "bypass," that is, in a side stream taken off the system and leading back to the system rather than directly in-line.

C

Calcareous Coating or Deposit. A layer consisting of a mixture of calcium carbonate and magnesium hydroxide deposited on surfaces being cathodically protected against corrosion, because of increased pH adjustment to the protected surface.

Calcium Chloride. A substance used to obtain calcium chloride brine.

Calcium Sulfate. Chemical compound ($CaSO_4$) which is used at a drying agent or desiccant in liquid line dryers.

Calcium. A scale forming element found in boiler feedwater.

Calibration. A process of dividing and numbering the scale of an instrument; also of correcting or determining the error of an existing scale, or of evaluating one quantity in terms of readings of another.

California Air Resources Board (CARB). A state regulatory agency charged with regulating the air quality in California.

Calorie. It is equal to the amount of heat required to raise the temperature of one gram of water one degree Celsius.

Calorimeter. Device used to measure quantities of heat or determine specific heats.

Cange of State. Change from one phase, such as solid, liquid or gas, to another.

Capacitance. The charge-storing ability of a capacitor, defined as the magnitude of charge stored on either plate divided by the applied voltage.

Capacitance. The property of an electric current that permits the storage of electrical energy in an electrostatic field and the release of that energy at a later time.

Capacitor, (Condenser). A device that can store an electric charge when voltage is applied.

Capacity. The adsorption activity possessed in varying degrees by ion-exchange materials. This quality may be expressed as kilograins per cubic foot, gram-milliequivalents per gram, pound-equivalents per pound, gram-milliequivalents per milliliter, and so on, where the numerators of these ratios represent the weight of the ions adsorbed and the denominators represent the weight or volume of the adsorbent.

Capillary Tube. The capillary tube is a metering device made from a thin tube approximately 0.5 to 6 metre long and from 0.025 to 0.090 inches in diameter which feeds liquid directly to the evaporator. Usually limited to systems of 1 ton or less, it performs all of the functions of the thermal expansion valve when properly sized.

Capillary. The name given to the thin tube attached to the bulb which transmits the bulb pressure changes to the controller or indicator. The cross sectional area of the capillary is extremely small compared to the cross section of the bulb so that the capillary, which is usually outside of the controlled fluid, will introduce the smallest possible error in the signal being transmitted from the bulb.

Carbon Dioxide (CO_2). A colorless, odorless, non-toxic gas. It is one of main products of fossil-fuel combustion. Carbon dioxide is a greenhouse gas that contributes to the potential for global warming.

Carbon Dioxide. Compound of carbon and oxygen (CO_2) which is sometimes used as a refrigerant. Refrigerant number is R-744.

Carbon Filter. Air filter using activated carbon as air cleansing agent.

Carbon Monoxide (CO). A colorless, odorless and toxic gas. It blocks the lungs' ability to obtain oxygen. CO is produced by incomplete combustion of fossil fuels and is a major part of air pollution. Compression ignition (diesel) engines generate significantly lower CO emissions than spark ignited engines.

Carbon Rings. To minimize steam leakage, gland sealing arrangement features an angle-type, carbon ring design with a packing case that is an integral part of the turbine case. Inconel springs hold the rings in place.

Carbon Steel. Steel which owes its properties chiefly to various percentages of carbon without substantial amounts of other alloying elements; also known as ordinary, straight carbon, or plain carbon steel.

Carbon Tetrachloride. Colorless nonflammable and very toxic liquid used as a solvent. It should never be allowed to touch skin and fumes must not be inhaled.

Carbonaceous Exchanger. Ion-exchange materials of limited capacity prepared by the sulfonation of coal, lignite, peat, and so on.

Carbonate Hardness. That hardness caused in water by bicarbonates and carbonates of calcium, and magnesium.

Carbonate. An ion or salt of carbonic acid, containing carbon and oxygen such as calcium carbonate. (CaC03)

Carbonate-Polymer Treatment. A treatment method using synthetic polymers, generally used with high hardness (60-70) ppm and high alkalinity.

Carboxylic. A term describing a specific acidic group (COOH) that contributes cation-exchange ability to some resins.

Carcinogens. Substances known to cause cancer.

Carryover. The moisture and entrained solids forming the film of steam bubbles, as a result of foaming in a boiler. This condition is caused by a faulty boiler water condition. See also foaming.

Cascade Systems. Arrangement in which two or more refrigerating systems are used in series; uses evaporator of one machine to cool condenser of other machine. Produces ultra-low temperatures.

Cascade. A series of stages in which the output of one stage is the input of the next stage.

Cast Iron. A ferrous alloy; the carbon content is greater than the maximum solubility in austenite at the eutectic temperature.

Casting. Generic term referring to a process where a fluid material (usually a molten alloy) is made to flow into a shaped mold cavity where it solidifies; this method is used to produce complex component shapes and properties difficult to achieve otherwise.

Catalyst. A substance which influences the rate of a chemical reaction but is not one of the original reactants or final products, i.e. it is not consumed or altered in the reaction. Catalysts are used in many processes in the chemical and petroleum industries. Emission control catalysts are used to promote reactions that change exhaust pollutants from internal combustion engines into harmless substances.

Cathode. In electrolysis or electrochemical corrosion, a site on a surface where actions in solution are neutralized by electrons to become elements that either plate out on the surface or react with water to produce a secondary reaction.

Cathodic Protection. A method of preventing corrosion by making the metal a cathode in a conducting medium by means of a direct electrical current that is galvanic.

Cathodic Protection. Reduction of corrosion rate by shifting the corrosion potential of the electrode towards less oxidizing potential by applying an external electromotive force.

Cation. A positively charged ion that migrates through the electrolyte toward the cathode under the influence of a potential gradient.

Cation-Exchange Softeners. Ion exchange units are known primarily as water softeners. But they can also remove nitrates, sulfates, and various toxic metals from water. ion exchangers.pdf

Cationic. The condition of a polymer, colloid, or large particle having exchangeable anions on its surface and an opposite, positive charge on the substrata.

Caustic Cracking. A form of stress-corrosion cracking most frequently encountered in carbon steels or iron-chromium-nickel alloys that are exposed to concentrated hydroxide solutions at temperature of 200 to 250□C.

Caustic Embrittlement. An obsolete term replaced by caustic cracking.

Caustic Soda. A common water treatment chemical, sodium hydroxide.

Cavitation Erosion. Progressive loss of original material from a solid surface due to continuing exposure to cavitation.

Cavitation. The formation and collapse, within a liquid, of cavities or bubbles that contain vapor or gas or both. In general, cavitation originates from decreases in static pressure in the liquid. In order to erode a solid surface by cavitation, it is necessary for the cavitation bubbles to collapse on or close to that surface.

Celsius Temperature Scale. A thermometric scale in which the freezing point of water is called 0□C and its boiling point 100□C at normal atmospheric pressure.

Cementite. Iron carbide.

Centane Number. A measure of ignition quality of a fuel or petroleum with reference to normal centane high-ignition quality fuel with an arbitrary number of 100.

Centrifugal Compressor. Pump which compresses gaseous fluids by centrifugal force.

Centrifugal Force. 1. Is actually not a force but the experience of an inertial force experienced in a rotating reference frame acting away from the center of the rotation. It is equal in magnitude but opposite to the centripetal force required to constrain the body to move in a circular motion. s actually not a force but the experience of an inertial.

2. On a centrifugal pump, it is that force which throws water from a spinning impeller.

Centrifugal Pump. 1. A pump consisting of an impeller fixed on a rotating shaft and enclosed in a casing, having an inlet and a discharge connection. The rotating impeller creates pressure in the liquid by the velocity derived from centrifugal force.

2. Pump which produces fluid velocity and converts it to pressure head.

Centripetal Force. Is the force that causes an object to move in a circle, acting towards the centre of the circle. In the case of a satellite the centripetal force is gravity, in the case of an object at the end of a rope, the centripetal force is the tension of the rope.

Ceramic. A compound of metallic and nonmetallic elements, for which the interatomic bonding is predominantly ionic.

Cetane Index. A calculated value, derived from fuel density and volatility, giving a reasonably close approximation to cetane number.

Cetane Number. A measure of ignition quality of diesel fuel. The higher the cetane number the easier the fuel ignites when injected into an engine. Cetane number is determined by an engine test using two reference fuel blends of known cetane numbers. The reference fuels are prepared by blending normal cetane (n-hexadecane), having a value of 100, with heptamethyl nonane, having a value of 15.

CFR (Cooperative Fuel Research) Engine. A single cylinder, overhead valve, variable compression ratio engine used for measuring octane or cetane quality.

Change of State. Condition in which a substance changes from a solid to a liquid or a liquid to a gas caused by the addition of heat. Or the reverse, in which a substance changes from a gas to a liquid, or a liquid to a solid, caused by the removal of heat.

Channeling. Cleavage and furrowing of the bed due to faulty operational procedure, in which the solution being treated follows the path of least resistance, runs through these furrows, and fails to contact active groups in other parts of the bed, (water treatment).

Charge. Amount of refrigerant placed in a refrigerating unit.

Charging Board. Specially designed panel or cabinet fitted with gauges, valves and refrigerant cylinders used for charging refrigerant and oil into refrigerating mechanisms.

Charpy Test. An impact test in which a V-notched, keyhole-notched, or U-notched specimen, supported at both ends horizontally, is struck behind the notch by a striker mounted at the lower end of a pendulum; the energy that is absorbed in fracture is calculated from the height to which the striker would have risen had there been no specimen and the height to which it actually rises after fracture of the specimen.

Check Valve. Device which permits fluid flow in one direction.

Chelate. Is a molecule, similar to an ion exchanger, capable to withdraw ions from their water solutions into soluble complexes.

Chemical Cleaning. Using a solvent solution to remove mill scale and corrosion products.

Chemical Feedline. The line which feeds the boiler treatment chemicals into the boiler.

Chemical Precipitation. When the chemicals react with the dissolved minerals in the water to produce a relative insoluble reaction product. A typical example of this takes place with the lime-soda softening process.

Chemical Refrigeration. System of cooling using a disposable refrigerant. Also called an expendable refrigerant system.

Chemical Stability. Resistance to chemical change which ion-exchange resins must possess despite contact with aggressive solutions.

Chill Factor. Calculated number based on temperature and wind velocity.

Chilled-Water System. A re-circulating water system using water chilled in a refrigeration machine as a source for cooling.

Chiller/Heaters. A unit that supplies either chilled water for cooling or hot water for heating, (HVAC).

Chloride. An ion, compound, or salt of chlorine, such as sodium chloride (NaCl) or calcium chloride ($CaCl_2$).

Chlorination. A process in which chlorine gas or other chlorine compounds are added to the water for the purpose of disinfecting.

Choke Tube. Throttling device used to maintain correct pressure difference between high-side and low-side in refrigerating mechanism. Capillary tubes are sometimes called choke tubes.

Chordal Thermocouple. A thermocouple installed in furnace tubes, designed to measure the effectiveness of water treatment within the boiler.

Circuit Breaker. A switch-type mechanism that opens automatically when it senses an overload (excess current).

Circuit. 1. An electrical arrangement requiring a source of voltage, a closed loop of wiring, an electric load and some means for opening and closing it. 2. Interconnections of electrical elements and devices that perform a desired electrical function.

Clay. Finely suspended earth mineral sometimes found as an impurity in water.

Clean Air Act (CAA). In the U.S., the fundamental legislation to control air pollution. The original Clean Air Act was signed in 1963. The law set emissions standards for stationary sources, such as factories and power plants. Criteria pollutants included lead, ozone, CO, SO_2, NOx and PM, as well as air toxics. The CAA was amended several times, most recently in 1990. The Amendments of 1970 introduced motor vehicle emission standards for automobiles and trucks.

Clean-Fuel Vehicle (CFV). A vehicle that has been certified to meet clean-fuel standards of the Clean Air Act Amendments of 1990.

Clearance Space. Space between top of piston and the valve plate.

Clearance Vapor. The vapor remaining in the clearance space at the end of each discharge stroke.

Cleavage. Transcrystalline fracture along specific crystallographic planes; usually associated with low-energy fracture; may exhibit river patterns and/or tongues.

Closed Cycle. is the gas turbine arrangement, in which the exhaust is directed back again to compressor without coming in contact with the atmospheric air.

Closed Feed Water Heater. An indirect-contact feed water heater. Steam and water are separated by tubes.

Closed Re-Circulating Water System. A system using as a heat-transfer medium water that continuously circulates through closed piping and heat exchanger without evaporation.

Cloud Point (CP). A measure of the ability of a diesel fuel to operate under cold weather conditions. Defined as the temperature at which wax first becomes visible when diesel fuel is cooled under standardized test conditions (ASTM D2500).

CO Coagulant. A substance that promotes the clumping of particulate matter in water, forming a larger mass and thus promoting settling of particulates and clarification of the water.

CO. Carbon monoxide (CO) is a colorless, odorless, poisonous gas. It is produced by the incomplete burning of solid, liquid, and gaseous fuels. Appliances fueled with natural gas, liquified petroleum (LP gas), oil, kerosene, coal, or wood may produce CO. Burning charcoal produces CO. Running cars produce CO.

CO2. is an atmospheric gas composed of one carbon and two oxygen atoms. Carbon dioxide results from the combustion of organic matter if sufficient amounts of oxygen are present. It is also produced by various microorganisms in fermentation and is breathed out by animals. Plants absorb carbon dioxide during photosynthesis, using both the carbon and the oxygen to construct carbohydrates. It is present in the Earth's atmosphere at a low concentration and acts as a greenhouse gas. It is a major component of the carbon cycle.

Coagulation. Is the process whereby finely divided particles of turbidity and color, capable of remaining in suspension indefinitely, are combined by chemical means into masses sufficiently large to effect rapid settling.

Coalescence. The gathering together of coagulated colloidal liquids into a single continuous phase.

Code Installation. Refrigeration or air conditioning installation which conforms to the local code and/or the national code for safe and efficient installations.

Co-Efficient of Conductivity. Measure of the relative rate at which different materials conduct heat. Copper is a good conductor of heat and, therefore, has a high coefficient of conductivity.

Coefficient of Expansion. A measure of the change in length or volume of an object, specifically, a change measured by the increase in length or volume of an object per unit length or volume.

Coefficient of Friction. The dimensionless ratio of the friction force (F) between two bodies to the normal force (N) pressing these bodies together - D(f) = (F/N)

Coefficient of Performance (COP). Ratio of work performed or accomplished as compared to the energy used under designated operating conditions.

Coefficient of Thermal Expansion. The fractional change in length (or sometimes in volume, when specified) of a material for a unit change in temperature.

Co-Generation Generation. a term applied to identify the generation of people interested in co-generation.

Cogeneration. A term used to describe the combination of different thermodynamic cycles for the purpose of increasing all-over cycle efficiency.

Cold Deck. The cooling section of a mixed air zoning system.

Cold Filter Plugging Point (CFPP). A measure of the ability of a diesel fuel to operate under cold weather conditions. Defined as the lowest temperature at which diesel fuel will pass through a fine wire mesh screen of the test apparatus.

Cold Junction. That part of a thermoelectric system which absorbs heat as the system operates.

Cold Process. A water treatment process carried out at room temperature.

Cold Wall. Refrigerator construction which has the inner lining of refrigerator serving as the cooling surface.

Cold Working. Plastic deformation of a metal at a temperature below that at which it recrystallizes; increasing the amount of cold work causes the dislocation density to rise in the material, making it more difficult to plastically deform the material and eventually cause brittle fracture.

Cold. The absence of heat; a temperature considerably below normal.

Colloidal. A state of suspension in a liquid medium in which extremely small particles are suspended and dispersed but not dissolved.

Colloids. Organic matter of very fine particle size, usually in the range of 10^{-5} to 10^{-7} cm in diameter. It tends to inhibit the formation of dense scale and results in the deposition of sludge, or causes it to remain in suspension, so that it may be blown from the boiler.

Column Operation. Conventional utilization of ion-exchange resins in columns through which pass, either upflow or down flow, the solution to be treated.

Columnar Structure. Coarse structure of parallel columns of grains caused by highly directional solidification of molten metal resulting from sharp thermal gradients.

Combined Feeder Cutoff. A device that regulates makeup water to a boiler in combination with a low-water fuel cutoff.

Combined Steam-Gas Plant. where a gas turbine is combined with steam plant in order to utilize the waste heat.

Combined Treatment. A method of physical treatment , followed by the addition of chemicals to remove oxygen.

Combustion. The act or process of burning.

Comfort Chart. A chart showing effective temperatures with dry-bulb temperatures and humidities (and sometimes air motion) by which the effects of various air conditions on human comfort may be compared.

Comfort Cooler. System used to reduce the temperature in the living space in homes. These systems are not complete air conditioners as they do not provide complete control of heating, humidifying, dehumidification, and air circulation.

Comfort Cooling. Refrigeration for comfort as opposed to refrigeration for storage or manufacture.

Comfort Zone. (Average) the range of effective temperatures over which the majority (50 percent or more) of adults feels comfortable; (extreme) the range of effective temperatures over which one or more adults feel comfortable. An area on the psychrometric chart which shows conditions of temperature, humidity and sometimes air movement in which most people are comfortable.

Common Neutral. A neutral conductor that is common to, or serves, more than one circuit.

Common Rail Injection. A diesel fuel injection system employing a common pressure accumulator, called the rail, which is mounted along the engine block. The rail is fed by a high pressure fuel pump. The injectors, which are fed from the common rail, are activated by solenoid valves. The solenoid valves and the fuel pump are electronically controlled. In the common rail injection system the injection pressure is independent from engine speed and load. Therefore, the injection parameters can be freely controlled. Usually a pilot injection is introduced, which allows for reductions in engine noise and NOx emissions.

Component. An element or chemical compound that helps make up a material system; the composition of a phase or system can be described by giving the relative amounts of each component.

Composition. The elements or chemical components that make up a material and their relative proportions.

Compound Gauge. Instrument for measuring pressures both above and below atmospheric pressure.

Compound Refrigerating Systems. System which has several compressors or compressor cylinders in series. The system is used to pump low pressure vapors to condensing pressures.

Compound. They are chemically combined elements with definite proportions of the component elements.

Compressed Natural Gas (CNG). Natural gas compressed to a volume and density that is practical as a portable fuel supply.

Compressibility. The ease which a fluid may be reduced in volume by the application of pressure, depends upon the state of the fluid as well as the type of fluid itself.

Compression Ignition (CI). The form of ignition that initiates combustion in a diesel engine. The rapid compression of air within the cylinders generates the heat required to ignite the fuel as it is injected.

Compression Ratio. Ratio of the volume of the clearance space to the total volume of the cylinder. In refrigeration it is also used as the ratio of the absolute low-side pressure to the absolute high-side pressure.

Compression, Adiabatic. Is compressing a gas without removing or adding heat.

Compression. Term used to denote increase of pressure on a fluid by using mechanical energy.

Compressor Displacement. Volume, in cubic inches, represented by the area of the compressor piston head or heads multiplied by the length of the stroke.

Compressor Seal. Leak proof seal between crankshaft and compressor body in open type compressors.

Compressor Surging. An instability of air flow with axial compressor on the first stages of these compressors. Air flow might even be reversed that point.

Compressor Turbine. in terms of a gas turbine arrangement, it is the turbine which drives the compressor only.

Compressor, Clearance Pocket. Small space in a cylinder from which compressed gas is not completely expelled. This space is called the compressor clearance space or pocket. For effective operation, compressors are designed to have as small a clearance space as possible.

Compressor, Rotary Blade. Mechanism for pumping fluid by revolving blades inside cylindrical housing.

Compressor, Single-Stage. Compressor having only one compressive step between low-side pressure and high-side pressure.

Compressor. 1. Pump of a refrigerating mechanism which draws a low pressure on cooling side of refrigerant cycle and squeezes or compresses the gas into the high-pressure or condensing side of the cycle.

2. The pump which provides the pressure differential to cause fluid to flow and in the pumping process increases pressure of the refrigerant to the high side condition. The compressor is the separation between low side and high side.

Condensate Polisher. A device used to clean the returning condensate to the boiler feedwater system.

Condensate Pump. Device to remove water condensate that collects beneath an evaporator.

Condensate. The liquid formed by condensation of a vapor. In steam heating, water condensed from steam; in air conditioning, water extracted from air, as by condensation on the cooling coil of a refrigeration machine.

Condensation Polymerization. The formation of polymer macromolecules by an intermolecular reaction involving at least two monomer species, usually with the production of a by-product of low molecular weight, such as water.

Condensation. Process of changing a vapor into liquid by extracting heat. Condensation of steam or water vapor is effected in either steam condensers or dehumidifying coils, and the resulting water is called condensate.

Condense. Action of changing a gas or vapor to a liquid.

Condenser Tube. The heat transfer surface in a condenser.

Condenser. An apparatus used to transfer heat from a hot gas, simultaneously reducing that gas to a liquid.

Condenser-Water System. A re-circulating cooling water used as a heat transfer fluid for the condensation of a gas.

Condensing Temperature. The temperature at which the condensing gas is returned to a liquid at the same pressure.

Condensing Turbine. With this turbine, the steam exhausts to the condenser and the latent heat of the steam is transferred to the cooling water. The condensed steam is returned to the boiler as feedwater.

Condensing Unit Service Valves. Shutoff valves mounted on condensing unit to enable service technicians to install and/or service unit.

Condensing Unit, Refrigerant. An assembly of refrigerating components designed to compress and liquefy a specific refrigerant, consisting of one or more refrigerant compressors, refrigerant condensers, liquid receivers (when required) and regularly furnished accessories.

Condensing Unit. Part of a refrigerating mechanism which pumps vaporized refrigerant from the evaporator, compresses it, liquefies it in the condenser and returns it to the refrigerant control.

Conductance, Electrical. The reciprocal (opposite) of resistance and is the current carrying ability of any wire or electrical component. Resistance is the ability to oppose the flow of current.

Conductance, Surface Film. Time rate of heat flow per unit area under steady conditions between a surface and a fluid for unit temperature difference between the surface and fluid.

Conduction. Transfer of heat by direct contact.

Conductivity (Electrical). The ability of a liquid to conduct an electrical current and indicating the presence of cations and anions. Conductivity is usually expressed in Micromohs per cm.

Conductivity (Thermal). The time rate of heat flow through unit thickness of an infinite slab of homogeneous material in a direction perpendicular to the surface, induced by unit temperature difference. (W/m ? K)

Conductivity Meter. An electric instrument used to measure the conductivity of water to determine its content of dissolved solids.

Conductivity, Electrical. The proportionality constant between current density and applied electric field; a measure of the ease with which a material is capable of conducting an electric current.

Conductivity, Thermal. The time rate of heat flow through unit area and unit thickness of a homogeneous material under steady conditions when a unit temperature gradient is maintained in the direction perpendicular to area. Materials are considered homogeneous when the value of the thermal conductivity is not affected by variation in thickness or in size

Conductor. Substance or body capable of transmitting electricity or heat.

Conduit. A round cross-section electrical raceway, of metal or plastic.

Conformal Coating. A thin nonconducting coating that is either plastic or inorganic; it is applied to a circuit for environmental and mechanical protection.

Congealer. Also known as freezer.

Congruent Phosphate Control. Similar as a coordinated phosphate control but more restrictive where the equilibrium is based on maintaining a ratio of 2.6 Na/1.0 PO_4, instead of 3.0/1.0 PO_4.

Connected Load. The sum of all loads on a circuit. (1) Connection in Parallel: System whereby flow is divided among two or more channels from a common starting point or header. (2) Connection in Series: System whereby flow through two or more channels is in a single path entering each succeeding channel only after leaving the first or previous channel.

Constrictor. Tube or orifice used to restrict flow of a gas or a liquid.

Contamination. The introduction into water of microorganisms, chemicals, toxic materials, waste water in a concentration that makes the water unfit for its next intended use.

Control Point. The value of the controlled variable which the controller operates to maintain.

Control Valve. Valve which regulates the flow or pressure of a medium which affects a controlled process. Control valves are operated by remote signals from independent devices using any of a number of control media such as pneumatic, electric or electrohydraulic.

Control. A device for regulation of a system or component in normal operation, manual or automatic. If automatic, the implication is that it is responsive to changes of pressure, temperature or other property whose magnitude is to be regulated.

Controlled Device. One which receives the converted signal from the transmission system and translates it into the appropriate action in the environmental system. For example: a valve opens or closes to regulate fluid flow in the system.

Controller. A device capable of measuring and regulating by receiving a signal from a sensing device, comparing this data with a desired value and issuing signals for corrective action.

Convection, Forced. Convection resulting from forced circulation of a fluid, as by a fan, jet or pump.

Convection, Natural. Circulation of gas or liquid (usually air or water) due to differences in density resulting from temperature changes.

Convection. The movement of a mass of fluid (liquid or gas) caused by differences in density in different parts of the fluid; the differences in density are caused by differences in temperature. As the fluid moves, it carries with it its contained heat energy, which is then transferred from one part of the fluid to another and from the fluid to the surroundings.

Convergent Nozzle. Impulse Turbine - Stationary convergent nozzles are used for smaller pressure drops where the minimum exit pressure is 0.577 x the inlet pressure (the critical pressure for nozzles.) If the exit pressure is less than 0.577 x inlet pressure, eddy-currents are developed and the exit velocity will be less than calculated.

Convergent-Divergent Nozzle. Impulse Turbine - Stationary convergent-divergent nozzles prevent eddy-currents and the calculated velocity will be obtained even at large pressure drops.

Converted Vehicle. A vehicle, originally designed to operated on gasoline or diesel, that has been modified or altered to operate on an alternative fuel.

Cooler. Heat exchanger which removes heat from a substance.

Cooling Effect, Sensible. The difference between the total cooling effect and the dehumidifying effect, usually in watts.

Cooling Effect, Total. Difference between the total enthalpy of the dry air and water vapor mixture entering the cooler per hour and the total enthalpy of the dry air and water vapor mixture leave air the cooler per hour, expressed in watts.

Cooling Tower. Device for lowering the temperature of water by evaporative cooling, in which water is showered through a space through which outside air circulates. A portion of the water evaporates, its latent heat of vaporization cooling that portion of the water which does not evaporate.

Cooling, Evaporative. Involves the adiabatic exchange of heat between air and water spray or wetted surface. The water assumes the wet-bulb temperature of the air, which remains constant during its traverse of the exchanger.

Cooling, Regenerative. Process of utilizing heat which must be rejected or absorbed in one part of the cycle to function usefully in another part of the cycle by heat transfer.

Coordinated Phosphate Control. A treatment to prevent caustic gauging. Free caustic is eliminated by maintaining an equilibrium between the sodium and phosphate. Control is based on maintaining a ratio of 3.0 Na to/1.0 PO_4.

Coordination Number. The number of atomic or ionic nearest neighbors.

Cope. The top half of a horizontally parted mold.

Copolymer. A polymer that consists of two or more dissimilar mer units in combination along its molecular chains.

Cordierite. A ceramic material of the formula $2MgO\text{-}2Al_2O_3\text{-}5SiO_2$ which is used for automotive flow-through catalyst substrates and ceramic wall-flow diesel filters.

Corrator. A device or probe employed to measure current flow in a process flow. It consists of two identical electrodes, to which a small current is applied and measured, from which corrosion rates can be calculated.

Corrosion, Anode. The dissolution of an metal acting as an anode.

Corrosion, Atmospheric. The gradual degradation or alteration of a material by contact with substances present in the atmosphere, such as oxygen, carbon dioxide, water vapor, and sulfur and chlorine compounds.

Corrosion, Biological. Deterioration of metals as a result of the metabolic action of microorganisms. Also often named fouling.

Corrosion, Cathodic. Corrosion resulting resulting from a cathodic condition of a structure usually caused by the reaction of an amphoteric metal with the alkaline products of electrolysis.

Corrosion, Cavitation. A process involving conjoint corrosion and cavitation.

Corrosion, Concentration Attack. A form of corrosion caused by the concentration of caustic or phosphate salts under porous deposits, generally iron oxide. Sometimes found at welded tubes and due to steam blanketing.

Corrosion, Concentration-Cell. Pitting type of corrosion caused by an electrical potential differential between surfaces of a metal as a result of deposits or differences in the solution concentration in contact with the metal.

Corrosion, Coupons. Pre-weighed metal strips installed into fluid systems for the purpose of monitoring metal losses.

Corrosion, Crevice. Localized corrosion of a metal surface at, or immediately adjacent to an area that is shielded from full exposure to the environment because of close proximity between metal and the surface of another metal.

Corrosion, Deactivasion. The process of prior removal of the active corrosive constituents, usually oxygen, from a corrosive liquid by controlled corrosion of expendable metal or by other chemical means, therby making the liquids less corrosive.

Corrosion, Deposit (also called poultice corrosion). Corrosion occuring under or around a discontinous deposit on a metallic surface.

Corrosion, Effect. A change in any part of the corrosion system caused by corrosion.

Corrosion, Electrochemical. Corrosion that is accompanied by a low of electrons between cathodic and anodic areas on metallic surfaces.

Corrosion, Embrittlement. The severe loss of ductility of a metal resulting from corrosive attack, usually intergranular and often not visible.

Corrosion, External. A chemical deterioration of the metal on the fireside of boiler heating surfaces.

Corrosion, Fatigue. The process in which a metal fractures prematurely under conditions of simultaneous corrosion and repeated cyclic loading at lower stress levels or fewer cycles than would be required in the absence of the corrosive environment.

Corrosion, Filiform. Corrosion that occurs under organic coatings on metals as fine wavy hairlines.

Corrosion, Fretting. A type of corrosion which occurs where metals slide over each other. Long tubes in heat exchangers often vibrate, causing metal to metal contact, tube supports etc.. The metal to metal rubbing causes mechanical damage to the protective oxide coating.

Corrosion, Galvanic. Corrosion of a metal caused by its contact with a metal of lower activity; this contact results in an electron flow or current and dissolution of one of the metals.

Corrosion, Gaseous. Corrosion with gas as the only corrosive agent and without any aqueous phase on the surface of the metal. Also called dry corrosion.

Corrosion, General. A form of deterioration that is distributed more or less uniformly over a surface.

Corrosion, Graphitic. Corrosion of grey iron in which the iron matrix is selectively leached away, leaving a porous mass of graphite behind. This type of corrosion occurs in relativel mild aquous solutions and on buried piping.

Corrosion, Hot. An accelerated corrosion of metal surfaces that results from the combined effect of oxidation and ractions with sulfur compounds or other contminants such as chlorides, to form a molten salt on a metal surface that fluxes, destroys or disrupts the normal protective oxide.

Corrosion, Impingement. A form of erosion-corrosion generally associated with local impingement of a high velocity, flowing fluid against a solid surface.

Corrosion, Influenced. The corrosion cause by organisms due to their discharge containing sulfur compounds and the depolarization with other types of discharge due to the presence of the microorganisms.

Corrosion, Inhibitors. Substances that slow the rate of corrosion.

Corrosion, Intercrystaline. (See Intergranular Cracking).

Corrosion, Intergranular. Localized attack occurring on the metal grain boundaries. This is commonly found with stainless steels which have been improperly heat treated.

Corrosion, Internal. Usually refers to the internal corrosion and is considered an electrochemical deterioration of the boiler surface at or below the water surface.

Corrosion, Localized. 1. Corrosion at discrete sites, for example, crevice corrosion, pitting, and stress-corrosion cracking. 2. Non-uniform corrosion of a metal surface highlighted by spotty or pitting-type corrosion.

Corrosion, Microbial. (See biological corrosion).

Corrosion, Oxygen Deficiency. A form of crevice corrosion in which galvanic corrosion proceeds because oxygen is prevented from diffusing into the crevice.

Corrosion, Potential. The voltage between a corroding metal and a reference electrode.

Corrosion, Poultice. (See corrosion, deposit).

Corrosion, Poultice. A term used in the automotive industry to describe the corrosion of vehicle body parts due to the collection of road salts and debries on ledges and in pockets that are kept moist by weather and washing.

Corrosion, Protection. Modification of a corrosion system so that corrosion damage is mitigated.

Corrosion, Resistance. The ability of a material to resist deterioration by chemical or electrochemical reaction with its environment.

Corrosion, Stray Current. A form of attack caused by electrical currents going through unintentional path.

Corrosion, Stress Corrosion Cracking. Material deterioration due to cracking, by being under static stress either applied or residual.

Corrosion, Stress. Preferential attack of areas under stress in a corrosive environment, where such a environment alone would not have caused corrosion.

Corrosion, Sweet. The deterioration of metal caused by contact with carbon dioxide in water.

Corrosion, Thermo-Galvanic. Corrosion resulting from an electrochemical cell caused by a thermal gradient.

Corrosion, Uniform. The simplest form of corrosion. It attacks all surfaces exposed to a corrodent.

Corrosion. The chemical or electrochemical reaction between a material, usually a metal, and its environment that produces a deterioration of the material and its properties.

Corrosive Wear. A material deterioration due to the co-joint action of corrosion and mechanical action.

Corrosivity. An indication of the corrosiveness of the water of material. The corrosivity of a water as described by the water's pH, alkalinity, hardness, temperature, total dissolved solids, dissolved oxygen concentration and the Langerier Index.

Coulombic Force. A force between charged particles, such as ions.

Counter-Flow Heat Exchanger. When the fluid to be cooled flows against the direction of the coolant. In heat exchange between two fluids, opposite direction of flow, coldest portion of one meeting coldest portion of the other.

Covalent Bond. A bond in which two atoms share pair of electrons.

Covalent Bond. A primary interatomic bond that is formed by the sharing of electrons between neighboring atoms.

Crankshaft Seal. Leak proof joint between crankshaft and compressor body.

Craze Cracking (Or Checking). Irregular surface cracking of metal associated with thermal cycling.

Creep Strength. The constant nominal stress that will cause a specified quantity of creep in a given time at constant temperature. Creep strength is expressed as the stress necessary to produce 0.1% strain in 1000 hours.

Creep, Dynamic. Creep that occurs under conditions of fluctuating load or fluctuating temperatures.

Creep. The time-dependent permanent deformation that occurs under stress; for most materials it is important only at elevated temperatures.

Creep. Time dependent permanent strain under stress. This is used to rate the resistance of a material to plastic deformation under sustained load.

Crisper. Drawer or compartment in refrigerator designed to provide high humidity along with low temperature to keep vegetables, especially leafy vegetables - cold and crisp.

Critical Humidity. The relative humidity above which the atmospheric corrosion rate of some metals increase sharply.

Critical Point. A point at which the saturated liquid and saturated vapor states are identical. Also, the latent heat of evaporization is zero at this point.

Critical Pressure. The pressure at the critical temperature above which the fluid no longer has the properties of a liquid, regardless of further increase of pressure.

Critical Resolved Shear Stress. The shear stress, resolved within a slip plane and direction, which is required to initiate slip.

Critical Temperature. That temperature above which the vapor phase cannot be condensed to liquid by an increase in pressure.

Critical Velocity. Maximum velocity is obtained and no further pressure drop is obtained. (Acoustic Velocity).

Cross Compound Turbine. These turbines are large turbines with parallel shafts with a generator on each shaft. The steam flows through the high pressure turbine, then is crossed-over to the low pressure turbine

Crosslinked Polymer. A polymer in which adjacent linear molecular chains are joined at various positions by covalent bonds.

CRT. Cathode ray tube terminal.

Cryogenic Fluid. Substance which exists as a liquid or gas at ultra-low temperatures - 157□C.

Cryogenic Superconductor System. Uses helium to cool conductors to within few degrees of absolute zero where they offer no electric resistance.

Cryogenics. Refrigeration which deals with producing temperatures of -157□C below zero and lower.

Crystal Formation, Zone of Maximum. Temperature range in freezing in which most freezing takes place, i.e., about 25¡F to 30¡F for water.

Crystal Structure. For crystalline materials, the manner in which atoms or ions are arrayed in space; it is defined in terms of the unit cell geometry and the atom positions within the unit cell.

Crystalline. The state of a solid material characterized by a periodic and repeating three-dimensional array of atoms, ions, or molecules.

Crystallites. Atoms arranged in a repeating and definite structure.

Crystallization. Act or process of forming crystals or bodies by elements or compounds solidifying so they are bounded by plane surfaces, symmetrically arranged, and are external expressions of definite internal structure.

Crystallization. The separation, usually from a liquid phase on cooling, of a solid crystalline phase.

Cure. To irreversibly polymerize a thermosetting plastic by subjecting it to a time-temperature profile.

Curie Temperature. The temperature above which a ferromagnetic or ferrimagnetic material becomes paramagnetic.

Current (I). The electric flow in an electric circuit, which is expressed in amperes (amps).

Current Density. The current flowing to or from a unit area of an electrode surface.

Curtis Method or Curtis Stage. Velocity Compounding - This turbine design consists of one set of nozzles in which the steam is expanded from initial to exhaust pressure. The velocity of the steam resulting from this expansion is absorbed in two or more rows of moving blades. Rows of fixed or guide blades, attached to the casing, are set between rows of moving blades and receive and redirect the steam to the next row of moving blades. As the velocity is absorbed in more than one row of moving blades, the blade speed is less than if the velocity was all absorbed in one row of blades.

Cycle, Reversible. Theoretical thermodynamic cycle, composed of a series of reversible processes, which can be completely reversed.

Cycle, Water Treatment. A complete course of ion-exchange operation. For instance, a complete cycle of cation exchange would involve regeneration of the resin with acid, rinse to remove excess acid, exhaustion, backwash, and finally regeneration.

Cycle. A series of thermodynamic processes during which the working fluid can be made to undergo changes involving energy transition and is subsequently returned to its original state.

Cycles Of Concentration. The number of times the soluble mineral salts in a water supply have been concentrated in, a system.

Cycles. A system that undergoes a series of processes and always returns to its initial state.

Cylinder Head. Plate or cap which encloses compression end of compressor cylinder.

D

Dalton's Law of Partial Pressure. Each constituent of a mixture of gases behaves thermodynamically as if it alone occupied the space. The sum of the individual pressures of the constituents equals the total pressure of the mixture.

Damper. A device used to vary the volume of air passing through an air outlet, air inlet or duct.

Dashpot. A damping device, usually consisting of a cylinder and a piston in which relative motion of either displaces a fluid such as air or oil, resulting in friction.

Deadband. In HVAC, a temperature range in which neither heating nor cooling is turned on; in load management, a kilowatt range in which loads are neither shed nor restored.

De-Aerating Heaters. Mechanical device using steam to strip dissolve gases from the boiler feedwater and heating the feedwater.

De-Aeration. Act of separating air from substances.

De-Aerator. An apparatus or device which is used to remove dissolved air or oxygen from water.

De-Alkalization. The removal of alkalinity from a water supply by neutralization or ion exchange.

De-Alkalizer. An apparatus or device used to remove the alkaline carbonate and bicarbonate ions from a water supply.

De-Alloying. This is a corrosion process whereby one constituent of a metal alloy is preferentially removed from the alloy, leaving an altered residual microstructure.

Deashing. The removal from a solution of inorganic salts by means of adsorption by ion-exchange resins of both the cations and the anions that comprise the salts. See deionization.

De-Carbonation. Refers to the removal of carbon dioxide from the boiler feedwater.

Decibel (dB). A decibel is a division of a logarithmic scale for expressing the ratio of two quantities proportional to power or energy. The number of decibels denoting such a ratio is ten times the logarithm of the

Deconcentrator. This is a cylindrical tank connected before the boiler to receive the boiler feedwater before entering the boiler. It is designed to promote settling of suspended solids, which then could be removed via its own blowdown device. Was used for operation with very high suspended solids.

Dedicated Vehicle. A vehicle designed to operate solely on one alternative fuel, as opposed to a converted vehicle which was later altered to use an alternative fuel.

Defect Structure. Relating to the kinds and concentrations of vacancies and interstitials in a ceramic compound.

De-Flocculant. An electrolyte adsorbed on colloidal particles in suspension that charges the particles to create repulsion forces which maintain the particles in a dispersed state, thus reducing the viscocity of the suspension.

Defrost Cycle. Refrigerating cycle in which evaporator frost and ice accumulation is melted.

Defrost Timer. Device connected into electrical circuit which shuts unit off long enough to permit ice and frost accumulation on evaporator to melt.

Defrosting Control. Device to automatically defrost evaporator. It may operate by means of a clock, door cycling mechanism or during "off" portion of refrigerating cycle.

Defrosting Type Evaporator. Evaporator operating at such temperatures that ice and frost on surface melts during off part of operating cycle.

Defrosting. Process of removing frost accumulation from evaporators.

Degree Day. A unit, based upon temperature difference and time, used in estimating fuel consumption and specifying nominal heating load of a building in winter. For any one day, when the mean temperature is less than 65°F there exist as many degree days as there are Fahrenheit degrees difference in temperature between the mean temperature for the day and 65°F.

Degree of Polymerization. The average number of mer units per polymer chain molecule.

Degrees of Superheat. The amount by which the temperature of a superheated vapor exceeds the temperature of the saturated vapor at the same pressure.

Dehumidification. The condensation of water vapor from air by cooling below the dewpoint or removal of water vapor from air by chemical or physical methods.

Dehumidifier. (1) An air cooler or washer used for lowering the moisture content of the air passing through it; (2) An absorption or adsorption device for removing moisture from air.

Dehumidifying. Removal of moisture from the air.

Dehydration. The removal of water vapor from air, stored goods or refrigerants.

Deionization. Deionization, a more general term than deashing, embraces the removal of all charged constituents or ionizable salts (both inorganic and organic) from solution.

De-Ionizer. An apparatus or device used to remove the ions of dissolved salts from water.

Delta Iron. An allotropic (polymorphic) form of iron stable above 2550□ F and crystallizing in the body centered cubic lattice.

Demand Charge. That part of an electric bill based on kW demand and the demand interval, expressed in dollars per kilowatt. Demand charges offset construction and maintenance of a utility's need for a large generating capacity.

Demand Control. A device which controls the kW demand level by shedding loads when the kW demand exceeds a predetermined set point.

Demand Interval. The period of time during which kW demand is monitored by a utility service, usually 15 or 30 minutes long.

Demand Load. The actual amount of load on a circuit at any time. The sum of all the loads which are ON. Equal to the connected load minus the loads that are OFF.

Demand Reading. Highest or maximum demand for electricity an individual customer registers in a given interval, example, 15 minute interval. The metered demand reading sets the demand charge for the month.

Demand. The probable maximum rate of water flow as determined by the number of water supply fixture units.

Demineralizer. A process to remove dissolved matter from boiler pretreated water by contacting the water with ion-exchange resins.

Density, Absolute. Mass per unit volume of a solid material, expressed usually in kg/m^3.

Density. The ratio of the mass of a specimen of a substance to the volume of the specimen. The mass of a unit volume of a substance. When weight can be used without confusion, as synonymous with mass, density is the weight per unit volume.

Desalination. The removal of inorganic dissolved solids from water.

Desicant, Liquid. A hygroscopic liquid, such as glycol, used to remove water from other fluids.

Desiccant. Any absorbent or adsorbent, liquid or solid, that will remove water or water vapor from a material. In a refrigeration circuit, the desiccant should be insoluble in the refrigerant.

Design Pressure. 1. Highest or most severe pressure expected during operation. Sometimes used as the calculated operating pressure plus an allowance for safety.

2. Highest or most severe pressure expected during operation. Sometimes used as the calculated operating pressure plus an allowance for safety.

Design Working Pressure. The maximum allowable working pressure for which a specific part of a system is designed.

Desilicizer. An apparatus or device used to remove silica from a water supply.

Dessert Bag. A canvas bag which permits seepage of its liquid. The liquid will evaporate and obtains the to evaporate partly from the content of the bag and thus cooling its content.

Detergent Additive. In lubrication technology, a surface active additive that helps to keep solid particles suspended in an oil

Detergent Cleaning. A boiler cleaning process using an alkaline solution, primarily to remove oil and grease.

Detergent Oil. A heavy duty oil containing a detergent additive. These oils are mainly used in combustion engines.

Detergent-Dispersant. A compound mixture of cleaning agents that have both surface-active properties and suspending properties.

Dew Point Depression. The difference between dry bulb and dew point temperatures.

Dew Point Temperature. The temperature at which condensation begins, if air is cooled at constant pressure.

Dew Point. Temperature at which vapor (at 100 percent humidity) begins to condense and deposit as liquid.

Diamagnetism. A weak form of induced or nonpermanent magnetism for which the magnetic susceptibility is negative.

Diaphragm. Flexible material usually made of thin metal, rubber or plastic.

Diatomaceous Earth Filtration. Is a process in which a filter cake or precoat of diatomaceous earth is used as a filter medium.

Die. (1) a metal form used as a permanent mold for die casting or for a wax pattern in investment casting; (2) an integrated circuit chip as diced or cut from the finished wafer.

Dielectric Constant. A relative measurement of the degree of polarization (shift of positive charge toward the negative electrode and negative charge toward the positive electrode) that occurs when a material is placed in an electric field.

Dielectric Fitting. A non conductive substance such as plastic that is placed between two dissimilar metals to prevent galvanic current flow.

Dielectric Strength. A measure of the ability of a dielectric (insulator) to withstand a potential difference across it without electric discharge.

Dielectric. 1. A nonconductor of electricity. 2. Any material that is electrically insulating.

Diesel Oxidation Catalyst (DOC). Catalyst promoting oxidation processes in diesel exhaust. Usually designed to reduce emissions of the organic fraction of diesel particulates, gas-phase hydrocarbons, and carbon monoxide.

Diesel Particulate Filter (DPF). A device which physically captures diesel particulates preventing their discharge from the tailpipe. Collected particulates need to be removed from the filter, usually by continuous or periodic oxidation in a process called "regeneration".

Diesel Particulate Matter (DPM). Sub-micron size particles found in diesel exhaust. Most emission regulations specify DPM measurement methods in which particulates are sampled on filters from cooled exhaust gas. The cooling causes condensation of vapors in the gas sampling train. Thus, the DPM is composed of both solid and liquid particles and is generally classified into three fractions: (1) inorganic carbon (soot), (2) organic fraction (often referred to as SOF or VOF), and (3) sulfate fraction (hydrated sulfuric acid).

Differential Aeration Cell. An electrolytic cell, the electomagnetic force of which is due to a difference in air (oxygen) concentration at one electrode as compared with that at another electrode of the same material. (See concentration cell)

Differential Solute Concentration. A potential difference between an anode and cathode on metal, because of a concentration cell due to dissolved metals.

Differential. The temperature or pressure difference between cut-in and cut-out temperature or pressure of a control.

Diffraction (x-ray). Constructive interference of x-ray beams that are scattered by atoms of a crystal.

Diffuser. A circular, square, or rectangular air distribution outlet, generally located in the ceiling and com prised of deflecting members discharging supply air in various directions and planes, and arranged to promote mixing of primary air with secondary room air.

Diffusion Coefficient (*D*). The constant of proportionality between the diffusion flux and the concentration gradient in Fick's first law; its magnitude is indicative of the rate of atomic diffusion.

Diffusion Flux (*J*). The quantity of mass diffusing through and perpendicular to a unit cross-sectional area of material per unit time.

Diffusion. Motion of atoms, ions, or vacancies through a material.

Dimethyl Ether (DME). The simplest ether CH_3-O-CH_3. Can be manufactured from natural gas or from a renewable organic feedstock. DME is a prospective alternative diesel fuel.

Dipole. A system or object whose one end has a negative charge and the other a positive charge.

Direct Acting. Instruments that increase control pressure as the controlled variable (such as temperature or pressure) increases; while reverse acting instruments increase control pressure as the controlled variable decreases.

Direct Current. A source of power for an electrical circuit which does not reverse the polarity of its charge.

Direct Injection (DI). In diesel engines with direct injection the combustion chamber is not divided and fuel is injected directly to the cylinder.

Direct-Expansion Evaporator. One that contains only enough liquid to continue boiling as heat is absorbed by it.

Dislocation Line. The line that extends along the end of the extra half-plane of atoms for an edge dislocation, and along the center of the spiral of a screw dislocation.

Dislocation. A linear crystalline defect around which there is atomic misalignment; plastic deformation corresponds to the motion of dislocations in response to an applied shear stress; edge, screw, and mixed dislocations are possible.

Dispersant Oil. A heavy duty oil containing a dispersant additive.

Dispersant. A chemical which causes particulates in a water system to remain in suspension.

Displacement Pump. Pumps in which energy is added to the water periodically and the *water* is contained in a set volume.

Displacement Volume. The volume displaced by the piston between top dead center and bottom dead center.

Dissociation. Ionization.

Dissolved Gases. Gases soluble in water.

Dissolved Solids (TDS). The measure of the total amount of dissolved matter.

Distillation. Involves boiling water and condensing the vapor.

DMA. Direct memory access. A process where block of data can be transferred between main memory and secondary memory without processor intervention.

DOD. The U.S. Department of Defense.

DOE. The U.S. Department of Energy.

Domain. A volume region of a ferromagnetic or ferrimagnetic material in which all atomic or ionic magnetic moments are aligned in the same direction.

Domestic Hot Water. Potable hot water as distinguished from hot water used for house heating.

Doping. The intentional alloying of semiconducting materials with controlled concentrations of donor or acceptor impurities.

DOT. The U.S. Department of Transportation.

Dowel Pins. Metal pins that guarantee exact alignment of a flange joint on a horizontally split casing as found in turbine construction

Downflow. Conventional direction of solutions to be processed in ion-exchange column operation, that is, in at the top, out at the bottom of the column.

Downstream. The outlet side of an instrument, a pump, valve, etc..

Draft Gauge. Instrument used to measure air movement by measuring air pressure differences.

Drag. The bottom half of a horizontally parted mold.

Drier. Substance or device used to remove moisture from a refrigeration system.

Drierite. Desiccant which operates by chemical action.

Drift. 1. Entrained water in the stack discharge of a cooling tower. 2. Term used to describe the difference between the set point and the actual operating or control point.

Droop. Terms used to describe the difference between the set point and the actual operating or control point.

Drum Water Level Line. The water level in the drum during the normal operating mode.

Dry Bulb Temperature. The temperature registered by an ordinary thermometer. The dry bulb temperature represents the measure of sensible heat, or the intensity of heat.

Dry Bulb. An instrument with a sensitive element to measure ambient air temperature.

Dry Compression. The compression of vapor, in a vapor-liquid vapor-compression refrigeration cycle.

Dry Ice. Refrigerating substance made of solid carbon dioxide which changes directly from a solid to a gas (sublimates). Its subliming temperature is -78□C.

Dry Pipe. A perforated or slotted pipe or box inside the drum and connected to the steam outlet.

Dry Standby. A method of sealing al water and steam connections and placing a desiccant in the unit and applying an airtight seal.

Dry System. Refrigeration system which has the evaporator liquid refrigerant mainly in the atomized or droplet condition.

Dryness Fraction Or Quality. Weight fraction of the vapor in a vapor-liquid mixture.

Dual Shaft Gas Turbine. a gas turbine which has one turbine on one shaft driving the compressor and when the gas discharged from this turbine is directed to another turbine on a separate shaft to drive a load.

Dual-Fuel Vehicle. A vehicle designed to operate on a combination of alternative fuel, such as compressed natural gas (CNG) or liquefied petroleum gas (LPG), and conventional fuel, such as diesel or gasoline. These vehicles have two separate fuel systems, which inject both fuels simultaneously into the engine combustion chamber.

Dual-Temperature Re-Circulating Water System. A closed re-circulating water system that uses water either for cooling, by circulating it through a chiller, or for heating, by circulating it through a boiler or heat-exchanger depending upon need.

Ductile Fracture. A mode of fracture that is attended by extensive gross plastic deformation.

Ductile Gouging. Referring to irregular wasting of the tube metal beneath a porous deposit The micro structure of the metal does not change with this process and the ductility remains, but the thinning leads to rapture.

Ductile-to-brittle Transition. The transition from ductile to brittle behavior with a decrease in temperature exhibited by BCC alloys; the temperature range over which the transition occurs is determined by Charpy and Izod impact tests.

Ductility. A measure of a material's ability to undergo appreciable plastic deformation before fracture; it may be expressed as percent elongation or percent area reduction from a tensile test.

Ductility. The ability of a material to deform plastically without fracturing.

Dummy Piston. The axial thrust in reaction turbines can be nearly eliminated by the use of balance or dummy pistons. With the correct size of a dummy piston exposed to two different bleed point pressures, the thrust is nearly equalized. There is a small leakage across the labyrinth seal of the dummy piston as steam leaks from the high to the lower bleed point

Dust. An air suspension (aerosol) or particles of any solid material, usually with particle size less than 100 microns.

Dynamic Discahrge Head. Static discharge head plus friction head plus velocity head.

Dynamic Load. An imposed force that is in motion, that is, one that may vary in magnitude, sense, and direction.

Dynamic Pumps. Pumps in which energy is added to the water continuously and the water is not contained in a set volume.

Dynamic Suction Head. Positive static suction head minus friction head and minus velocity head.

Dynamic Suction Lift. The sum of suction lift and velocity head at the pump suction when the source is below pump centerline.

Dynamic System. An ion-exchange operation, wherein a flow of the solution to be treated is involved.

E

Economizer. A series of tubes located in the path of flue gases. Feedwater is pumped through these tubes on its way to the boiler in order to absorb waste heat from the flue gas.

Eddy Current Testing. An electromagnetic nondestructive testing method in which eddy-current flow is induced in the test object. Changes in flow caused by variations in the object are deflected into a nearby coil or coils where they are measured.

Eddy Currents (Electricity). An electromagnetic nondestructive testing method in which eddy-current flow is induced in the test object. Changes in flow caused by variations in the object are deflected into a nearby coil or coils where they are measured.

Eddy Currents (Steam Turbines). Impulse Turbine - As the steam passes through convergent nozzles, if the exit pressure is less than 0.577 x inlet pressure (the critical pressure for nozzles), eddy-currents are developed and the exit velocity will be less than calculated.

Edge Dislocation. A linear crystalline defect associated with the lattice distortion produced in the vicinity of the end of an extra half-plane of atoms within a crystal; the Burgers vector is perpendicular to the dislocation line.

EDTA. A chelating agent used with boiler water treatment. Often referred as the replacement for the phosphate-hydroxide treatment method.

Effective Temperature. Overall effect on a human of air temperature, humidity and air movement.

Effluent. The solution which emerges from an ion-exchange column.

Elastic Deformation. Nonpermanent deformation, totally recovered upon release of an applied stress.

Elastomer. A polymeric material that may experience large and reversible elastic deformations.

Electric Defrosting. Use of electric resistance heating coils to melt ice and frost off evaporators during defrosting.

Electrical Circuit. A power supply, a load, and a path for current flow are the minimum requirements for an electrical circuit.

Electrochemical Reactions. A metal wasting process, due to the fluid (boiler water) being subjected to an electrical current.

Electrode Boiler. A boiler which generates steam or hot water by the action of immersed electrodes which conduct electricity through the boiler water, which, in turn, generates heat by its resistance to electric current.

Electrodeposition. (1) the deposition of a conductive material from a plating solution by the application of electrical current; (2) the deposition of a substance on an electrode by passing electric current through an electrolyte; electroplating, electroforming, electrorefining, and electrotwinning result from electrodeposition.

Electrodialysis. This is a membrane process where an applied electric charge draws impurity ions through permeable membranes to create high purity feedwater streams or low purity waste streams.

Electrolysis. Chemical decomposition caused by action of an electric current in a solution.

Electrolyte. 1. A chemical compound which dissociates or ionizes in water to produce a solution which will conduct an electric current; an acid, base, or salt.

2. A solution through which an electric current may be carried by the motion of ions.

Electromagnetic Radiation. Energy propagated at the speed of light by an electromagnetic field.

Electromechanical. Converting electrical input into mechanical action. A relay is an electromechanical switch.

Electronegative. For an atom, having a tendency to accept valence electrons; also a term to describe non-metallic elements.

Electronic Control Module (ECM). A microprocessor that determines the beginning and end of each injection cycle on every cylinder. The ECM determines both fuel metering and injection timing in response to such parameters as engine crankshaft position and rpm, engine coolant and intake air temperature, and absolute intake air boost pressure.

Electropositive. For an atom, having a tendency to release valence electrons; also a term to describe metallic elements.

Electroregeneration. Hydrogen and hydroxyl ions are formed be electrical splitting of water molecules and are swept through the unit by steady, low-voltage direct current, continuously cleansing the resin beads and carrying away the unwanted salts

Element. A pure substance that cannot be broken down by chemical means to a simpler substance.

Elemental Carbon (EC). Inorganic carbon, as opposed to carbon in organic compounds, sometimes used as a surrogate measure for diesel particulate matter, especially in occupational health environments. Elemental carbon usually accounts for 40-60% of the total DPM mass.

Elevation Head. The energy possessed per unit weight of a fluid because of its elevation.

Elution. The stripping of adsorbed ions from an ion-exchange material by the use of solutions containing other ions in concentrations higher than those of the ions to be stripped.

Emission Credit Trading. A program administered by the Environmental Protection Agency under which low polluters are awarded credits which may be traded on a regulated market and purchased by polluters who are in noncompliance for emissions until compliance can be achieved.

Emulsion. A colloidal dispersion of one liquid in another.

Endothermic Reaction. Pertaining to a chemical reaction which is accompanied by an absorption of heat.

Energy (Consumption) Charge. That part of an electric bill based on kWh consumption (expressed in cents per kWh). Energy charge covers cost of utility fuel, general operating costs, and part of the amortization of the utility's equipment.

Energy = power x time. (1) Expressed in kilowatt-hours (kWh) or watt hours (Wh), and is equal to the product of power and time. (2) The ability to do work. Energy can exist in one of several forms, such as heat, light, mechanical, electrical or chemical. Energy can neither be created nor destroyed, but can be transferred from one form to another. Energy can also exist in one of two states, either potential or kinetic.

3. In the simplest terms, energy is the ability to perform work. It may exist in several forms, such as heat energy, mechanical energy, chemical energy, or electrical energy, and may be changed from one form to another.

4. The ability to do work. Energy can exist in one of several forms, such as heat, light, mechanical, electrical or chemical. Energy can neither be created nor destroyed, but can be transferred from one form to another. Energy can also exist in one of two states, either potential or kinetic.

Engine Oil. An oil used to lubricate an internal combustion engine.

Engine. Prime mover; device for transforming fuel or heat energy into mechanical energy.

Enthalpy. The total quantity of heat energy contained in a substance, also called total heat; the thermodynamic property of a substance defined as the sum of its internal energy plus the quantity Pv/J, where P = pressure of the substance, v = its volume, and J = the mechanical equivalent of heat.

Entrainment (HVAC). The capture of part of the surrounding air by the air stream discharged from an outlet (some times called secondary air motion).

Entrainment. The transport of water into a gas stream. In a boiler, this is carryover, in a cooling tower, drift.

Entropy. The ratio of the heat added to a substance to the absolute temperature at which it is added.

Environment. The aggregate of all conditions (such as contamination, temperature, humidity, radiation, magnetic and electric fields, shock, vibration) that externally influence the performance of a material or component.

EPACT. The U.S. Energy Policy Act of 1992.

Epson Salt. Magnesium sulfate.

Equalizing Hole (Steam Turbine). A hole in the turbine disc designed to equalize axial thrust with impulse bladed turbines.

Equiaxed. Powder or grain shapes with approximately equal dimensions.

Equilibrium (phase). The state of a system where the phase characteristics remain constant over indefinite time periods; at equilibrium the free energy is a minimum.

Equilibrium Reactions. The interaction of ionizable compounds in which the products obtained tend to revert to the substance from which they were formed until a balance is reached in which both reactants and pacts are present in definite ratios.

Equivalent Weight. Refers to the amount of an element combining with a unit weight of hydrogen. In terms of water treatment, a method used to calculate the concentration of a given ion in terms of its calcium carbonate.

Erosion , Abrasive. Erosive wear caused by relative motion of solid particles which are present in fluids and are moving parallel to a solid surface.

Erosion, Cavitation. Progressive loss of original material from a solid surface due to continuing exposure to cavitation.

Erosion, Impingement. Loss of material from a solid surface due to liquid impingement.

Erosion, Liquid. Removal of films or metal by mechanical action and corrosion of active metal.

Erosion-Corrosion. A conjoint action involving corrosion and erosion in the presence of a moving corrosive fluid, leading to the accelerated loss of material.

Etching. Chemical surface corrosion, usually conducted in a controlled fashion on a polished surface of a material sample to reveal details of the microstructure.

Ethane (R-170). Refrigerant sometimes added to other refrigerants to improve oil circulation.

Eutectic Point. Freezing temperature for eutectic solutions.

Eutectic. An isothermal reversible reaction in which a liquid solution is converted into two or more intimately mixed solids on cooling.

Eutectic. Upon cooling, a liquid phase transforms isothermally and reversibly into two intimately mixed solid phases; the lowest melting composition in a material system.

Eutectoid. Upon cooling, one solid phase transforms isothermally and reversibly into two new solid phases that are intimately mixed.

EV. Electric vehicle.

Evacuation. The removal of gases from a system.

Evaporation. The change of state from liquid to vapor, for example as water evaporates to a vapor in a cooling tower.

Evaporative Condenser. 1. A condenser which has water flowing over coils containing the refrigerant gas which is thus cooled and condensed by evaporation of that water.

2. Device which uses open spray or spill water to cool a condenser. Evaporation of some of the water cools the condenser water and reduces water consumption.

Evaporative Cooling. The adiabatic exchange of heat between air and a water spray or wetted surface. The water approaches the wet-bulb temperature of the air, which remains constant during its traverse of the exchanger.

Evaporative Emissions. Hydrocarbon vapors that escape from a fuel storage tank or a vehicle fuel tank or vehicle fuel system.

Evaporator Fan. Fan which increases airflow over the heat exchange surface of evaporators.

Evaporator Pressure Regulator. Automatic pressure regulating valve mounted in suction line between evaporator outlet and compressor inlet. Its purpose is to maintain a predetermined pressure and temperature in the evaporator.

Evaporator, Flooded. Evaporator containing liquid refrigerant at all times.

Evaporator. The heat exchanger in which the medium being cooled, usually air or water, gives up heat to the refrigerant through the exchanger transfer surface. The liquid refrigerant boils into a gas in the process of the heat absorption.

Exflltration. The flow of air outward from a space through walls, leaks, etc.

Exfoliation. Scaling off of a surface in flakes or layers as the result of corrosion.

Exhaustion. The state in which the adsorbent is no longer capable of useful ion exchange; the depletion of the exchanger's supply of available ions. The exhaustion point is determined arbitrarily in terms of (1) a value in parts per million of ions in the effluent solution; and (2) the reduction inequality of the effluent water determined by conductivity bridge which measures the resistance of the water to the flow of an electric current.

Exothermic. Chemical reaction in which heat is released.

Expansion Joint. Device in piping designed to allow movement of the pipe caused by the pipe's expansion and contraction.

Expansion Tank. A reservoir usually above a closed re-circulating water system that is blanketed with a gas to permit expansion and contraction of water in the system during temperature changes.

Expansion Valve, Capillary Tube. A tube of small internal diameter used as liquid refrigerant flow control and pressure reducer between high and low sides. Also used to transmit pressure from the sensitive bulb of some temperature controls to the operating element.

Expansion Valve, Thermostatic. Control valve operated by temperature and pressure within evaporator. It controls flow of refrigerant. Control bulb is attached to outlet of evaporator.

Expansion Valve. Device in refrigerating system which reduces the pressure from the high side to the low side and is operated by pressure.

Expendable Refrigerant System. System, which discards the refrigerant after it has evaporated.

External Drive. Term used to indicate a compressor driven directly from the shaft or by a belt using an external motor. Compressor and motor are serviceable separately.

External Equalizer. Tube connected to low-pressure side of a thermostatic expansion valve diaphragm and to exit end of evaporator.

External Treatment. Refers to the treatment of water before it enters the boiler.

Extraction Turbine – are turbines where steam is extracted at one or more points at constant pressure. Extraction turbines may be single or double-extraction-condensing turbines or single- or double-extraction back-pressure turbines. The extracted steam is used for process. Do not call these bleed turbines where steam is used for heating feed water.

Extrusion. A forming technique whereby a material is forced, by compression, through a die orifice.

F

Face Area. The total plane area of the portion of a grille, coil, or other items bounded by a line tangent

Face-centered cubic (FCC). A crystal structure found in some of the common elemental metals; within the cubic unit cell atoms are located at all corner and face-centered positions.

Fahrenheit. A thermometric scale in which 32 (¡F) denotes freezing and 212 (¡F) the boiling point of water under normal pressure at sea level (14.696 psi).

Fail Safe. In load management, returning all loads to conventional control during a power failure. Accomplished by a relay whose contacts are normally closed.

Failure. A rupture, break, or disintegration of a metal or part of an HVAC system.

False Brineling. Damage to a solid bearing surface characterized by indentations not caused by plastic deformation resulting from overload, but thought to be due to other causes such as fretting corrosion.

Fan Performance Curve. Fan performance curve refers to the constant speed performance curve. This is a graphical presentation of static or total pressure and power input over a range of air volume flow rate at a stated inlet density and fan speed. It may include static and mechanical efficiency curves. The range of air volume flow rate which is covered generally ex tends from shutoff (zero air volume flow rate) to free delivery (zero fan static pressure). The pressure curves are generally referred to as the pressure-volume curves.

Fan Tube Axial. A propeller or disc type wheel within a cylinder and including driving mechanism supports for either belt drive or direct connection.

Fan, Centrifugal. A fan rotor or wheel within a scroll type housing and including driving mechanism sup ports for either belt drive or direct connection.

Fan, Propeller. A propeller or disc type wheel within a mounting ring or plate and including driving mechanism supports for either belt drive or direct connection.

Fan, Vaneaxial. A disc type wheel within a cylinder, a set of air guide vanes located either before or after the wheel and including driving mechanism supports for either belt drive or direct connection.

Farad. A unit of electric capacity, designated by *F*.

Fatigue. The phenomenon leading to fracture under repeated or fluctuating stresses having maximum value less than the ultimate strength of the material.

Fault. A short circuit either line to line, or line to ground.

Federal Test Procedure (FTP). Test cycle(s) used in the U.S. for emission testing and certification of engines and vehicles. The chassis dynamometer cycle for light duty vehicle testing is commonly referred to as *FTP-75*. The engine dynamometer cycle for testing of heavy-duty (HD) engines is known as *HD FTP*, or *FTP Transient* cycle.

Feed Water Heater. A device used to heat feed water with steam.

Feed Water Line. The piping leading to a system through which the feed water flows.

Feed Water. Water which is fed to a system such as a boiler or cooling tower.

Feeder. Part of the gating system that forms the reservoir of molten metal necessary to compensate for losses due to shrinkage as the metal solidifies; sometimes referred to as a riser.

Ferric Coagulant. Ferric sulfate $Fe_2(SO4)_3$ act to precipitate ferric hydroxide, coagulate at 4.0 - 11.0 pH range.

Ferric Hydroxide. The complete reaction product of iron, water, and oxygen, which forms a red precipitate in water [$Fe(OH)_3$]

Ferric Ion. An iron atom that has a positive electric charge of +3. (Fe^{2+})

Ferrimagnetism. Permanent and large magnetizations found in some ceramic materials; it results from antiparallel spin coupling and incomplete magnetic moment cancellation.

Ferrite (ceramic). Ceramic oxide materials composed of both divalent and trivalent cations, some of which are ferrimagnetic.

Ferrite (iron). Body-centered cubic iron; also iron and steel alloys that have the BCC crystal structure.

Ferroelectricity. Spontaneous alignment of electric dipoles within a material under the influence of an electric field, resulting in a hysteresis loop when the direction of electric field is switched.

Ferromagnetism. Spontaneous alignment of magnetic dipoles within a material under the influence of a magnetic field, resulting in a hysteresis loop when the direction of magnetic field is switched.

Ferrous Alloy. A metal alloy for which iron is the prime constituent.

Ferrous Hydroxide. The reaction product of iron and water in the absence of oxygen; it remains soluble in the water [$Fe(OH)_2$].

Ferrous Ion. An iron atom that has a positive electric charge of + 2(Fe^{2+}).

Ferrous. Metallic materials in which the principle component is iron.

Fick's First Law. The diffusion flux is proportional to the concentration gradient; this relationship is employed for steady-state diffusion situations.

Fick's Second Law. The time rate of change of concentration is proportional to the second derivative of concentration; this relationship is employed in nonsteady-state diffusion situations.

Acoustic methods vary widely in their ability to resolve reservoirs, as shown by this graph of the range versus resolution of the various techniques.

An air-gun array is towed several hundred meters behind a seismic vessel to provide an energy source for the acquisition of marine seismic data.

The angle between the normal to the interface of two media and an incident raypath is the angle of incidence, and is equal to the angle of reflection in isotropic media. The angle of refraction depends on the velocity of the wave in that medium. Snell's law describes the changes in the direction of a wavefront as it travels in media of different velocities.

Filming Amines. Amines that form a impervious non-wettable film, which acts as a barrier between the metal and the condensate and provide protection against carbon dioxide and oxygen. These amines do not neutralize carbon dioxide.

Filter. A device to remove solid material from a fluid.

Filter-Drier. A combination device used as a strainer and moisture remover.

Filtration. Is the process of passing a liquid containing suspended matter through a suitable porous material in such a manner as to effectively remove the suspended matter from the liquid.

Fin. An extended surface to increase the heat transfer area, as metal sheets attached to tubes.

Fire Point. The temperature at which a material will continue to burn for at least 5 seconds without the benefit of an outside flame.

Fire Tube. A tube, in a boiler, through which the hot gases flow and transfer heat to the water on the outside of the tube.

Fire Wall. The back end of a boiler, opposite the burner, at which the hot gases change direction of flow.

Firebrick. A refractory brick, often made from fire clay, that is able to withstand temperature in the range of 1500 to 1600□C, and is used to line furnaces.

Firing. A high temperature heat treatment that increases the density and strength of a ceramic piece.

Fixed Displacement Pump. A pump in which the displacement per cycle cannot be varied.

Flash Chamber. A separating tank placed between the expansion valve and the evaporator to separate and bypass any gas formed in the expansion valve.

Flash Gas. The gas resulting from the instantaneous evaporation of refrigerant in a pressure-reducing device to cool the refrigerant to the evaporating temperature obtaining at the reduced pressure.

Flash Point. The temperature at which a combustible liquid gives off just enough vapor to produce a vapor/air mixture that will ignite when a flame is applied. The flash point is measured in a standardized apparatus using standard test methods, such as ASTM D93 or ISO 2719.

Flash Point. The temperature at which a material to give off sufficient vapor to form a flammable mixture.

Flash Tank. A vessel used for separating the liquid phase from the gaseous phase formed from a rise in temperature and/or a reduction of pressure on the flowing stream.

Flash. The portion of a superheated fluid converted to vapor when its pressure is reduced.

Flashing. Evaporation of a liquid into a vapor.

Flexible-Fueled Vehicle. A vehicle with the ability to operate on alternative fuels, 100 percent petroleum-based fuels, or a mixture of alternative fuel and petroleum-based fuels.

Floating Action Controllers. Essentially two position type controllers which vary the position of the controlled devices but which are arranged to stop before reaching a maximum or minimum position.

Flocculants. An electrolyte added to a colloidal suspension to cause the particles to aggregate and settle out as the result of reduction in repulsion between particles.

Flocculation. The process of agglomerating coagulated particles into settable flocs, usually of a gelatinous nature.

Flood Back. The condition of liquid refrigerant returning, usually from an overfed evaporator, to the compressor through the suction line.

Flotation. A process of separating solids from water by developing a froth.

Flow Rate. The volume of solution which passes through a given quantity of resin within a given time. Flow rate is usually expressed in terms of feet per minute per cubic foot of resin or as milliliters per minute per milliliter of resin.

Flow, Laminar Or Streamline. Fluid flow in which each fluid particle moves in a smooth path substantially parallel to the paths followed by all other particles.

Flow, Turbulent. Fluid flow in which the fluid moves transversely as well as in the direction of the tube or pipe axis, as opposed to streamline or viscous flow.

Fluid Head. The static pressure of fluid expressed in terms of the height of a column of the fluid, or of some manometric fluid, which it would support.

Fluid. The general term that includes gas, vapor or liquid

Fluidized Bed. A contained mass of finely divided solid that behaves like a fluid when brought into suspension in a moving gas.

Flux. Chemically or physically active formulation capable of cleaning oxides and enabling wetting of metals with solder.

Fly Ash. A finely divided siliceous material formed during the combustion of coal, coke, or other solid fuels.

Foam Carryover. Is the development of excessive moisture in the steam from carryover of foam from the drum. Usually common in low pressure boilers due to high concentration of dissolved solids.

Foam. A polymer that has been made porous (or spongelike) by the incorporation of gas bubbles.

Foaming. Formation of steam bubbles on the surface of the boiler water due to high surface tension of the water.

Force Pump. A device used to inject a solution into a closed system through an opening such as a drain valve.

Force. The action on a body which tends to change its relative condition as to rest or motion.

Forced Convection. Movement of fluid by mechanical force such as fans or pumps.

Forced Draft Cooling Tower. Cools water by mechanically forcing air through the tower.

Force-Feed Oiling. Lubrication system which uses a pump to force oil to surfaces of moving parts.

Forging. Mechanical forming of a metal by heating and hammering.

Fossil Fuel. Any naturally occurring organic fuel formed in the Earth's crust, such as petroleum, coal and natural gas.

Fouling Factor. The degree of interference with heat transfer.

Fouling. Deposits of impurities, dirt or foreign matter that clog systems or restrict flow and interfere with heat transfer.

Free Energy. A thermodynamic quantity that is a function of both the internal energy and entropy (or randomness) of a system; at equilibrium the free energy is at a minimum.

Freeboard. The space provided above the resin bed in an ion-exchange column to allow for expansion of the bed during backwashing.

Freezer Burn. Condition applied to food which has not been properly wrapped and that has become hard, dry and discolored.

Freezer. A refrigerating device designed to lower the temperature below 0□C.

Freeze-Up. (1) Formation of ice in the refrigerant control device which may stop□the flow of refrigerant into the evaporator. (2) Frost formation on an evaporator which may stop the airflow through the evaporator.

Freezing Point. The temperature at which a liquid becomes solid.

Freezing. Change of state from liquid to solid.

Frenkel Defect. In an ionic solid, a cation-vacancy and cation-interstitial pair.

Freon. Trade name for a family of synthetic chemical refrigerants.

Frequency. The number of vibrations, waves, or cycles of any periodic phenomenon per second. In architectural acoustics, the interest lies in the audible frequency range of 20 to 20000 cps Hertz (cycles per second).

Fresh Water. Water that has little or no salt dissolved in it.

Friction Head. The pressure in psi or feet of the liquid pumped which represents system resistance that must be overcome.

Friction. Friction is the resistance found at the duct and piping walls. Resistance creates a static pressure loss in systems. The primary purpose of a fan or pump is to produce a design volume of fluid at a pressure equal to the frictional resistance of the system and the other dynamic pressure losses of the components.

Frigorific Mixture. Are substances used in laboratory methods of producing a drop in temperature. A common example is a mixture of snow and salt.

Frost Back. Condition in which liquid refrigerant flows from evaporator into suction line; usually indicated by sweating or frosting of the suction line.

Frost Control. Semiautomatic - Control which starts defrost part of a cycle manually and then returns system to normal operation automatically.

Frost Free Refrigerator. Refrigerated cabinet which operates with an automatic defrost during each cycle.

Frost. Frozen condensation.

Frosting Type Evaporator. Refrigerating system which maintains the evaporator at frosting temperatures during all phases of cycle.

Fuel cycle. The processes involved in extracting a fuel in its native form, converting it to a useful product, transporting it to market, and consuming it at its final destination.

Fuel Knock. A hammer like noise produced when fuel is not burned properly in a cylinder.

Full Load Current. See Running Current.

Fumes. Solid particles commonly formed by the condensation of vapors from normally solid materials such as molten metals. Fumes may also be formed by sublimation, distillation, calcination, or chemical reaction wherever such processes create airborne particles predominantly below one micron in size. Such solid particles sometimes serve as condensation nuclei for water vapor to form smog.

Fungus. A lower form of plant life which does not contain chlorophyll, for example, a mold.

Fusible Plug. Plug or fitting made with a metal of a known low melting temperature. Used as safety device to release pressures in case of fire.

G

Gage Pressure. Absolute pressure minus atmospheric pressure.

Gall. To damage the surface of a powder metallurgy compact or die part through adhesion of powder to the die cavity wall or punch surface.

Galling. (1) a condition whereby excessive friction between mating parts results in localized welding; subsequent spalling and a further roughening of the rubbing surfaces may follow; (2) a severe form of scuffing associated with gross damage to the surfaces or failure.

Galvanic Action. Wasting away of two unlike metals due to electrical current passing between them. The action is increased in the presence of moisture.

Galvanic Cell. (1) a cell in which chemical change is the source of electrical energy; it usually consists of two dissimilar conductors in contact with each other and with an electrolyte, or of two similar conductors in contact with each other and with dissimilar electrolytes; (2) a cell or system in which a spontaneous oxidation-reduction reaction occurs, the resulting flow of electrons being conducted in an external part of the circuit. (3) Electrolytic brought about by the difference in electric potential between two dissimilar metals.

Galvanic Corrosion. Corrosion associated with the current of a galvanic cell consisting of two dissimilar conductors in an electrolyte or two similar conductors in dissimilar electrolytes; where the two dissimilar metals are in contact, the resulting reaction is referred to as couple action.

Galvanic Couple. 1. A pair of dissimilar conductors, commonly metals, in electrical contact. 2. The connection of two dissimilar metals in an electrolyte that results in current flow through the circuit.

Galvanic Current. The electric current that flows between metals or conductive nonmetals in a galvanic couple.

Galvanize. To coat a metal surface with zinc using various processes.

Galvanizing. The coating of metal with another by an electrolytic process; for example, electrolytically zinc-coat steel is called galvanized steel.

Gamma Iron. The face-centered cubic form of pure iron, stable from 910 to 1400 C (1670 to 2550 F).

Gamma Ray. Short-wavelength electromagnetic radiation, similar to x-rays but of nuclear origin, with a range of wavelength from about 10^{-14} 10^{-10} m.

Gas Constant. The coefficient "R" in the perfect gas equation: PV = MRT.

Gas Lubrication. A system of lubrication in which the shape and relative motion of the sliding surfaces cause the formation of a gas film having sufficient pressure to separate the surfaces.

Gas Metal Arc Welding. An arc welding process that produces coalescence of metals by heating them with an arc between a continuous filler metal electrode and the workpieces; shielding is obtained entirely from an externally supplied gas.

Gas Refrigeration Cycle. Where the refrigerant remains in the gaseous phase throughout.

Gas Turbine Compressor. a compressor designed foe the use with gas turbine installations. This could be centrifugal or an axial compressor.

Gas Turbine Engine. A combustion engine utilizing a turbine, such as a jet engine.

Gas Turbine. 1. A turbine powered by a gas, such as the hot gas produced in combustion processes.

2. An engine in which gas , under pressure is formed by combustion, is directed against a series of turbine blades. The energy in the expanding gas is converted into rotary motion.

Gas Valve. Device in a pipeline for starting, stopping or regulating flow of gas.

Gas, Inert. A gas that neither experiences nor causes chemical reaction nor undergoes a change of state in a system or process; e.g., nitrogen or helium mixed with a volatile refrigerant.

Gas. 1. Usually a highly superheated vapor which, within acceptable limits of accuracy, satisfies the perfect gas laws. 2. Vapor phase or strata of a substance.

Gasification. When a substance is converted to become a gas.

Gasket. A device, usually made of a deformable material, that is used between two relatively static surfaces to prevent leakage.

Gate. The portion of the runner where the molten metal enters the mold cavity.

Gauge Manifold. Chamber device constructed to hold both compound and high-pressure gauges. Valves control flow of fluids through.

Gauge Vacuum. Instrument used to measure pressures below atmospheric pressure.

General Corrosion. Uniform overall corrosion of metal surfaces.

Generating Tube. A boiler tube used for evaporation.

Generator. A machine that changes that changes mechanical energy into electrical energy.

Genset. Generator set, a generating system comprising a combustion engine driving an electrical generator.

Geometric Surface Area (GSA). In monolith catalyst substrates, the total channel surface area per unit of substrate volume.

GFI, GFCI. Ground fault (circuit) interrupter - a device that senses ground faults and reacts by opening the circuit.

Gibbs Free Energy. Determines the relative stability of a material system at constant temperature and pressure; $G = H - TS$, where H is enthalpy, T is absolute temperature, and S is entropy of the system.

Glass transition temperature. The temperature at which, upon cooling, a noncrystalline ceramic or polymer transforms from a supercooled liquid into a rigid glass.

Glass. An amorphous material with three-dimensional primary atomic bonding.

Grain Boundaries. Referring to the junction of crystallites.

Grain. 1. A unit of weight; 0.0648 grams; 0.000143 pounds. 2. An individual crystal in a polycrystalline metal or ceramic.

Grains of Moisture. The unit of measurement of actual moisture contained in a sample of air. (7000 grains - one pound of water).

Grains Per Gallon. A unit of concentration. 1 gr/gal = 17.1 mg/L.

Gram. A unit of weight; 15.432 grains; 0.0022 pounds.

Gram-Milliquivalents. The equivalent weight in grams, divided by 1000.

Gravity, Specific. Density compared to density of standard material; reference usually to water or to air.

Gravity. The attraction exerted by the earth's mass on objects at its surface.

Grease, Block. A grease that is sufficiently hard to retain its shape in block or stick form.

Grease, Soda Based. A grease prepared from lubricating oil and sodium soap.

Grease. A lubricant composed of an oil thickened with a soap or other thickener to a solid or semisolid consistency.

Green Liquor. The liquor resulting from dissolved molten smelt from Kraft recovery furnace in water.

Greensand. Naturally occurring materials, composed primarily of complex silicates, which possess ion-exchange properties.

Griffith Crack Theory. A balance must be struck between the decrease in potential energy and the increase in surface energy resulting form the presence of a crack.

Grooving. A form of deterioration of boiler plate by a combination of localized corrosion and stress concentration.

Gross Vehicle Weight (GVW). The maximum loaded weight (including the vehicle itself, passengers, and cargo) for which a vehicle is designed, as specified by the manufacturer. Often used as a criterion of vehicle size for the purpose of legislation; the exact definition may vary depending on the jurisdiction.

Ground Bed. Cathodic protection, an interconnected group of impressed-current anodes that absorbs the damage caused by generated electric current

Ground Bus. A busbar in a panel or elsewhere, deliberately connected to ground.

Ground Coil. Heat exchanger buried in the ground. May be used either as an evaporator or as a condenser.

Ground Conductor. Conductor run in an electrical system, which is deliberately connected to the ground electrode. Purpose is to provide a ground point throughout the system. Insulation color green. Also called "green ground".

Ground Fault. An unintentional connection to ground.

Ground. Zero voltage, or any point connected to the earth or "ground".

Grout. To force sealing material into a soil, sand or confined small space; or the sealing material used in grouting.

Guinier-Preston (G-P) Zone. A small precipitation domain in a supersaturated metallic solid solution; it has no well-defined crystalline structure of its own and contains an abnormally high concentration of solute atoms; the formation of G-P zones constitutes the first stage of precipitation and is usually accompanied by a change in properties of the solid solution in which they occur.

H

H. **Hermetic.** Sealed so that the object is gas tight.

HAC. Hydrogen- assist cracking.

Halide Leak Detector. A device used to detect vapor leaks of halogen refrigerants. It uses acetylene as its base.

Halide Refrigerants. Family of refrigerants containing halogen chemicals.

Halide Torch. Type of torch used to safely detect halogen refrigerant leaks in system.

Halogens. Substance containing fluorine, chlorine, bromine and iodine.

Hard Water. Water that contains dissolved compounds of calcium, magnesium or both.

Hardenability. A measure of the depth to which a specific ferrous alloy may be hardened by the formation of martensite upon quenching from a temperature above the upper critical temperature.

Hardness Control. An action designed to remove hardness and at the same time to produce an equivalent amount of suspended solids.

Hardness of Calcium Carbonate. The expression ascribed to the value obtained when the hardness-forming salts are calculated in terms of equivalent quantities of calcium carbonate; a convenient method of reducing all salts to a common basic for comparison.

Hardness. 1. Are generally referred to the presence of calcium and magnesium content of the water.

2. The measure of a material's resistance to deformation by surface indentation or by abrasion.

3. The scale-forming and lather-inhibiting qualities which water, high in calcium and magnesium ions, possesses.

Head Dynamic or Total. In flowing fluid, the sum of the static and velocity heads at the point of measurement.

Head Pressure Control. Pressure-operated control which opens electrical circuit if high-side pressure becomes too high.

Head Pressure. Pressure which exists in condensing side of refrigerating system.

Head Static. The static pressure of fluid expressed in terms of the height of a column of the fluid, or of some manometric fluid, which it would support.

Head Velocity. Height of fluid equivalent to its velocity pressure in flowing fluid.

Head. 1. Pressure, usually expressed in feet of water, inches of mercury or millimeters of mercury. 2. The measure of the pressure of water expressed in feet of height of water: 1 psi = 2.31 feet of water.

Header. Length of pipe or vessel to which two or more pipe lines are joined carries fluid from a common source to various points of use.

Headloss. The loss of energy as a result of friction; commonly expressed in feet.

Heat Capacity. 1. The amount of heat necessary to raise the temperature of a given mass one degree. Numerically, the mass multiplied by the specific heat

2. The quantity of heat required to produce a unit temperature rise per mole of material.

Heat Conductor. A material capable of readily conducting heat. The opposite of an insulator or insulation.

Heat Engine. Mechanical devices which convert heat to work, such as the steam boiler, gas turbine, solar energy, refrigerators, steam engines, steam turbines.

Heat Exchanger. Device used to transfer heat from a warm or hot surface to a cold or cooler surface. (Evaporators and condensers are heat exchangers.)

Heat Lag. The time it takes for heat to travel through a substance heated on one side.

Heat Leakage. Flow of heat through a substance.

Heat of Compression. Mechanical energy of pressure changed into energy of heat.

Heat of Condensation. The latent heat given up by a substance as it changes from a gas to a liquid.

Heat of Fusion. The latent heat absorbed when a substance changes from a solid state to a liquid state.

Heat of Respiration. Process by which oxygen and carbohydrates are assimilated by a substance; also when carbon dioxide and water are given off by a substance.

Heat of Vaporization. The latent heat absorbed by a substance as it changes from a liquid to a vapor.

Heat Pipe. A refrigeration device with no moving parts, but containing a refrigerants.

Heat Pump. A device used to transfer heat from a low temperature to a high temperature medium also a reversed cycle in which work is the input and heat is rejected to a sink at a higher temperature than the source.

Heat Pump. A refrigerating system employed to transfer heat into a space or substance. The condenser provides the heat while the evaporator is arranged to pick up heat from air, water, etc. By shifting the flow of air or other fluid, a heat pump system may also be used to cool the space.

Heat Sink. Relatively cold surface capable of absorbing heat.

Heat Transfer. 1. Flow of heat by conduction, convection and radiation. 2. Movement of heat from one body or substance to another. Heat may be transferred by radiation, conduction, convection or a combination of these three methods.

Heat, Sensible. Heat which is associated with a change in temperature; specific heat exchange of temperature; in contrast to a heat interchange in which a change of state (latent heat) occurs.

Heat, Specific- The ratio of the quantity of heat required to raise the temperature of a given mass of any substance one degree to the quantity required to raise the temperature of an equal mass of a standard substance (usually water at 59 F) one degree.

Heat, Total (Enthalpy). The sum of sensible heat and latent heat between an arbitrary datum point and the temperature and state under consideration.

Heat. Form of energy which acts on substances to raise their temperature; energy associated with random motion of molecules.

Heat-Exchanger Materials. The metals or materials of construction of a heat exchanger.

Heating Coil. Heat transfer device consisting of a coil of piping, which releases heat.

Heating Control. Device which controls temperature of a heat transfer unit which releases heat.

Heating Surface. That surface which is exposed to the heating medium for absorption and transfer of heat to the medium.

Heat-Transfer Medium. The fluid, often water, which acts as the agent or medium in a heat exchanger through which heat is exchanged from one side to the other.

Henry. The unit of self-inductance or mutual inductance in the metric system. Its symbol is

Henry's Law. An expression for calculating the solubility of a gas in a fluid based on temperature and partial pressure.

Hermetic Compressor. Compressor which has the driving motor sealed inside the compressor housing. The motor operates in an atmosphere of the refrigerant.

Hermetic Motor. Compressor drive motor sealed within same casing which contains compressor.

Hermetic System. Refrigeration system which has a compressor driven by a motor contained in compressor dome or housing.

Hermetic. Sealing of an object so it is airtight.

Hertz. A unit in the metric system used to measure frequency in cycles per second. Its symbol is

Hexagonal Close-packed (HCP). The unit cell is of hexagonal geometry and is generated by the stacking of close-packed planes of atoms.

Hidden Demand Charge. Electric bill charges that are based on cents per kWh per kW demand contain a hidden demand charge. A low load factor for a building then penalizes the energy user through this "hidden" charge.

Hideout. Is the accumulation of chemicals on surfaces, in crevices or in deposits within the system during normal operation.

High Limit Control. A device which normally monitors the condition of the controlled medium and interrupts system operation if the monitored condition be comes excessive, for example a high level of fluid in a storage tank.

High Side. Parts of the refrigerating system subjected to condenser pressure or higher; the system from the compression side of the compressor through the condenser to the expansion point of the evaporator.

High-Pressure Cut-Out. Electrical control switch operated by the high-side pressure which automatically opens electrical circuit if too high pressure is reached.

High-Side Float. Refrigerant control mechanism which controls the level of the liquid refrigerant in the high-pressure side of mechanism.

High-Vacuum Pump. Mechanism which can create a vacuum in the 1000 to 1 micron range.

Hot Deck. The heating section of a multizone system. HOT GAS BYPASS. Piping system in refrigerating unit which moves hot refrigerant gas from condenser into low-pressure side.

Hot Gas Defrost. Defrosting system in which hot refrigerant gas from the high side is directed through evaporator for short period of time and at predetermined intervals in order to remove frost from evaporator.

Hot Isostatic Pressing. (1) a process for simultaneously heating and forming a compact in which the powder is contained in a sealed flexible sheet metal or glass enclosure and subjected to equal pressure from all directions at a temperature high enough to permit plastic deformation and sintering; (2) a process that subjects a component (casting, powder forgins, etc.) To both elevated temperature and isostatic gas pressure in an autoclave; simultaneous application of heat and pressure virtually eliminates internal voids and microporosity through a combination of plastic deformation, creep, and diffusion.

Hot Junction. That part of thermoelectric circuit which releases heat.

Hot Pressing. A method used to densify a material, whereby heat and pressure are applied simultaneously, and the pressure is typically applied unidirectionally via rigid tooling.

Hot Process Phosphate Softening. A process whereby the calcium and magnesium salts containing, constituting the hardness of water, are chemically precipitated and removed with phosphate in conjunction with caustic soda.

Hot Process. A water treatment process, when the water is heated above the room temperature.

Hot Well. A tank used to receive condensate from various sources on its passage back to the boiler through a feedwater system.

Hot Working. Any metal forming operation that is performed above a metal's recrystallization temperature.

Hot-Water Heating Boiler. A boiler in which no steam is generated and from which hot water is circulated for heating purposes and then returned to the boiler.

Hot-Water Re-Circulating System. A heating system using water as a heat-transfer medium through a heat exchanger or boiler to terminal heating unit.

HSC. Hydrogen stress cracking.

HSCC. Hydrogen- assisted stress-corrosion cracking.

Humidifier. A device to add moisture to air.

Humidifying Effect. The latent heat of vaporization of water at the average evaporating temperature times the weight of water evaporated per unit of time.

Humidifying. Adding of moisture to the air.

Humidistat. A regulatory device, actuated by changes in humidity, used for the automatic control of relative humidity.

Humidity Ratio. The ratio of the mass of the water vapor to the mass of dry air contained in the sample.

Humidity, Absolute. The weight of water vapor per unit volume.

Humidity, Percentage. The ratio of the specific humidity.

Humidity, Relative. The ratio of the mol fraction of water vapor present in the air, to the mol fraction of water vapor present in saturated air at the same temperature and barometric pressure; approximately, it equals the ratio of the partial pressure or density of the water vapor in the air, to the saturation pressure or density, respectively, of water vapor at the same temperature.

Humidity. Dampness of air.

Hunting. A surge of engine speed to higher number of revolutions per minute, followed by a drop to normal engines speed without manual movement of the throttle. Is often caused by a faulty or improperly adjusted governor.

HVAC Boiler. Boiler for heating or air conditioning (Absorption Refrigeration)

HVAC Equipment. Apparatus or equipment used in heating, ventilating, and air conditioning.

HVAC. Heating, Ventilating, and Air Conditioning.

Hybrid Electric Vehicle (HEV). Hybrid-electric vehicle. Various types of electric vehicles that use another power source to propel the vehicle or generate power for an electric drive train, or a combination of the two types.

Hydration. Absorption of water by a mineral that results in a change in the nature of the mineral.

Hydraulic Classification. The rearrangement of resin particles in an ion-exchange unit. As the backwash water flows up through the resin bed, the particles are placed in a mobile condition wherein the larger particles settle and the smaller particles rise to the top of the bed.

Hydraulic Coupling. A fluid connection between a prime mover and the machine it drives. It uses the action of liquid moving against blades to drive the machine.

Hydraulic Head. The force exerted by a column of liquid expressed by the height of the liquid above the point at which the pressure is measured. Although head refers to a distance or height, it is used to express pressure, since the force of the liquid column is directly proportional to its height. Also called head or hydrostatic head.

Hydraulic/Electronic Unit Injector (HEUI). A type of unit injector actuated by engine oil pressure rather than the camshaft. A very high oil pressure (up to 3,000 psi) is created by a separate oil pump. This high pressure is routed to every injector through a gallery. The engine's Electronic Control Module varies the pressure in response to engine speed and other parameters.

Hydrocarbons. Organic compounds containing only hydrogen and carbon atoms in various combinations.

Hydrogen Cooled Generator. High performance is provided by effective cooling and loss reduction.

Hydrogen Cycle. A complete course of cation-exchange operation in which the adsorbent is employed in the hydrogen or free acid form.

Hydrogen Damage. A type of corrosion occurring beneath a relative dense deposit. This type of damage can only occur if hydrogen was present in the metal. Failure takes place at thick edges in the form of fracture, rather then thinning. Hydrogen produces the corrosion reaction, moving into underlying metal, causing decarburization and intergranular fissuring of the structure. Damages usually result in larger pieces of metal being blown away, rather then just bursting.

Hydrogen Induced Cracking. Caused by the introduction of hydrogen during the welding process.

Hydrogen Ion. A portion of the molecule of water containing one atom of hydrogen which has a positive electric charge.

Hydrogen-Ion Concentration. The degree or quantity of hydrogen ions in a water solution.

Hydrologic Cycle. The cycle of water from evaporation through condensation to precipitation.

Hydrolysis. A chemical reaction between a mineral and water that results in dissolution of the mineral.

Hydromechanical Injection. An injection system in which mechanical parts work through hydraulic pressure to meter and time the injection of fuel. No electronics are incorporated into hydromechanical injection systems

Hydrometer. Floating instrument used to measure specific gravity of a liquid.

Hydronic System. A re-circulating water system used for heating and/or comfort cooling.

Hydrophilic. Having an affinity for water.

Hydrostatic Pressure. The pressure at any point in a liquid at rest; equal to the depth of the liquid multiplied by its density.

Hydroxyl. The term used to describe the anionic radical (OH-) which is responsible for the alkalinity of a solution.

Hygrometer. Instrument used to measure degree of moisture in the atmosphere.

Hygroscopic. Ability of a substance to absorb and release moisture and change physical dimensions as its moisture content changes.

Hypereutectoid Alloy. For an alloy system displaying a eutectoid, an alloy for which the concentration of solute is greater than the eutectoid composition.

Hypoeutectoid Alloy. For an alloy system displaying a eutectoid, an alloy for which the concentration of solute is less than the eutectoid composition.

Hysteresis (magnetic). The irreversible magnetic flux density-versus-magnetic field strength (*B*-versus-*H*) behavior found for ferromagnetic and ferrimagnetic materials; a closed *B*-*H* loop is formed upon field reversal.

Hz.**HIC.** Hydrogen-induced cracking. (Same as hydrogen embrittlement)

I

Ice Cream Cabinet. Commercial refrigerator which operates at approximately -18°C; used for storage of ice cream.

Ignition Delay. The length of time or number of degrees of crankshaft rotation between the beginning of injection and ignition of the fuel.

Ignition Quality. The ability of a fuel to ignite when it is injected into the compressed-air charge in a diesel cylinder. It is measured by an index called the cetane number.

Immiscible. Not capable of mixing (as oil and water).

Impedance (Z). The quantity in an AC circuit that is equivalent to resistance in a DC circuit, inasmuch as it relates current and voltage. It is composed of resistance plus a purely AC concept called reactance and is expressed, like resistance, in ohms.

Impeller. A rotating set of vanes designed to impart rotation to a mass of fluid.

Impingement. High-velocity flow of water or gas over a metal surface, causing premature failure by abrasion.

Inclusion. Foreign particle present as an undesirable impurity in a material.

In-Direct Injection (IDI). In diesel engines with in-direct injection the fuel is injected to an auxiliary pre-chamber. Combustion starts in the prechamber and propagates to the cylinder.

Inductance. The process when a second conductor is placed next to a conductor carrying AC current (but not touching it), the ever-changing magnetic field will induce a current in the second conductor.

Induction Heating. Heating by combined electrical resistance and hysteresis losses induced by subjecting a metal to varying magnetic field surrounding a coil carrying alternating current.

Induction. The capture of part of the ambient air by the jet action of the primary air stream discharging from a controlled device.

Inductive Loads. Loads whose voltage and current are out-of-phase. True power consumption for inductive loads is calculated by multiplying its voltage, current, and the power factor of the load.

Inductor. A fundamental element of electrical systems constructed of numerous turns of wire around a ferromagnetic or air core.

Inert Gas. A gas that does not readily enter into or cause chemical reactions.

Infiltration. Air flowing inward as through a wall, crack, etc.

Influent. The solution which enters an ion-exchange unit.

Infrared Reflow. Technique in which long wavelength light (IR) serves as the heat source to reflow solder and form solder joints.

Inherently Low Emission Vehicle (ILEV). Any vehicle that is certified to meet transitional low-emission vehicle standards established by the California Air Resources Board (CARB) and does not emit any evaporative emissions.

Inhibitor. An additive used to retard undesirable chemical action in a product. It is added in small quantities to gasoline's to prevent oxidation and gum formation, to lubricating oils to stop color change, and to corrosive environments to decrease corrosive action.

Injection Period. The time, measured in degrees of crankshaft rotation, between the beginning and end of injection. On engines with hydromechanical injection systems, it is controlled by the opening and closing of ports in the injector body or by the action of a plunger forcing fuel out of a cup. On electronic injection systems, it is determined directly or indirectly by the action of a solenoid valve.

In-Line Injection Pump. An injection pump with a separate cylinder and plunger for each engine cylinder. Each plunger is rotated by a rack to determine metering via ports in the body of the pump and helical cuts on the pump plungers. The plungers are driven off a camshaft, which usually incorporates a centrifugal or electronically controlled timing advance mechanism.

Inorganic Flux. An aqueous flux solution of inorganic acids and halides.

Inorganic Material. Are substances not derived from living things.

Inrush Current. The current that flows the instant after the switch controlling current flow to a load is closed. Also called "locked rotor current".

Inspection and Maintenance Programs (IM). In-use emission testing programs, primarily for passenger cars, used by some U.S. states to identify vehicles that cause excessive emissions which contribute to air pollution.

Instantaneous Rate. Method for determining when load shedding should occur. Actual energy usage is measured and compared to a present kilowatt level. If the actual kilowatt level exceeds a designated set point, loads will be shed until the actual rate drops below the set point.

Insulation, Thermal. Material which is a poor conductor of heat; used to retard or slow down flow of heat through wall or partition.

Insulator (electric). A nonmetallic material that has a filled valence band at 0 K and a relatively wide energy band gap; consequently the room-temperature electrical conductivity is very low.

Insulator. A material of such low electrical conductivity that a flow of current through it can usually be neglected. Similarly, a material of low thermal conductivity, such as that used to insulate structures.

Integrated Circuit. A microcircuit that consists of interconnected elements inseparably associated and formed in-situ on or within a single substrate, usually silicon, to perform an electronic circuit function.

Intercooled Cycle. Refers to a gas turbine employing two compressors. The compressed air from the first compressor is cooled before being discharged to second compressor.

Intercooling. Removal of heat from compressed gas between the compression stages.

Interdiffusion. Diffusion of atoms of one metal into another metal.

Intergranular Cracking. Cracking or fracturing that occurs between the grains or crystal in a polycrystalline aggregate. Also called intercrystalline cracking. Contrast with transgranular cracking.

Intergranular Fracture. Fracture of polycrystalline materials by crack propagation along grain boundaries.

Intermediate Solid Solution. A solid solution or phase having a composition range that does not extend to either of the pure components of the system.

Intermittent Blowdown. The blowdown is taken from the mud drum, waterwall headers or the lowest point of circulation.

Interstage Differential. In a multistage HVAC system, the change in temperature at the thermostat needed to turn additional heating or cooling equipment on.

Interstitial Site. Octahedral and tetrahedral open spaces within a close-packed arrangment of atoms or ions in which a cation can fit.

Interstitial Solid Solution. A solid solution wherein relatively small solute atoms occupy interstitial positions between the solvent or host atoms.

Intumescent Mat. Ceramic fiber mat which irreversibly expands after exposed to high temperature. Usually contains vermiculite. Intumescent mats are used in the canning of catalytic converters and diesel filters to hold the ceramic substrate inside the steel canister.

In-Use Deterioration (IUD). The effect of time and use on vehicle performance and emissions.

Investment Casting. This process is based on surrounding (investing) an expendable pattern, typically wax, with a ceramic mold and then removing (by melting or vaporizing) the pattern prior to pouring molten alloy into the mold; also known as lost wax and precision casting.

Ion Exchange Resin. Cross linked polymers that form salts within ions from aqueous solutions.

Ion Exchange. A reversible process by which ions are interchanged between solids and a liquid.

Ion. 1. An atom or radical in solution carrying an integral electric charge, either positive (cation) or negative (anion).

2. An atom with a positive charge because it has had electrons removed or a negative charge because it has had electrons added.

Ionic Bonding. A coulombic interatomic bond that exists between two adjacent and oppositely charged ions; one of the primary types of atomic bonding in ceramics.

Ionic Strength. A measure of strength of a solution based on both the concentrations and valences of the ions present.

Ionization. The process of separation of a molecule into its electrically charged atoms or parts.

Iron Bacteria. Are filementous organisms encountered in iron-bearing water.

Iron. A metallic element found as an impurity in water in very small amounts. Also a metal which is widely used in the construction of HVAC and plumbing equipment; the major component of steel.

Isentropic Process. A process carried out reversibly without energy interchange as heat. Also a processes carried out with no entropy change.

Isobaric or Isopiestic Process. A process carried out at constant pressure.

Isochoric Process or Isometric. A process during which the specific volume remains constant.

Isomerism. The phenomenon whereby two or more polymer molecules or mer units have the same composition but different structural arrangements and properties.

Isometric Process. A process carried out at constant volume.

Isotactic. A type of polymer chain configuration where all side groups are positioned on the same side of the chain molecule.

Isothermal Expansion and Contraction. Action which takes place without a temperature change.

Isothermal. At a constant temperature.

Isothermal. Changes of volume or pressure under conditions of constant temperature.

Isotropic. Having identical values of a property in all crystallographic directions.

Izod Test. A type of impact test in which a V-notched specimen, mounted vertically, is subjected to a sudden blow delivered by the weight at the end of a pendulum arm; the energy required to break off the free end is a measure of the impact strength or toughness of the material.

J

Jacking Oil Pump. provides oil supply when the turbine is placed on barring (slow rotation to stop the turbine from sagging or hogging)

Jet Compressor. A device employing a ventury tube so that a high pressure stream flowing through the nozzle creates a lower pressure or a vacuum into which the gas to be compressed flows. The gas is discharged from the nozzle with the expanded high-pressure medium.

Jominy End Quench Test Data. A plain carbon steel and an alloy steel were assessed using the Jominy End Quench test. The hardness of the samples was measured as a function of the distance from the quenched end to demonstrate the different hardenability of the two steels. The data is shown as Vickers and Rockwell hardness.

Joule. 1. English Scientist James Prescott Joule (1818 - 1889).

2. The unit used to measure heat, work, and energy in the metric system. Its symbol is *J*. It is the amount of energy required to move an object of 1 kg mass to a height of 1 m. Also called a newton-metre.

Joule-Thomson Effect. The change in gas temperature which occurs when the gas is expanded adiabatically from a higher pressure to a lower pressure. The effect for most gases, except hydrogen and helium, is a cooling of the gas.

Journal. That part of a shaft or axle that rotates relative to a radial bearing.

K

Kata Thermometer. Large-bulb alcohol thermometer used to measure air speed or atmospheric conditions by means of cooling effect.

Kelvin Scale (K). Thermometer scale on which unit of measurement equals the Celsius degree and according to which absolute zero is 0 degree, the equivalent of -273.16□C. Water freezes at 273.16 K. and boils at 373.16 K. The relationship - TK = TC + 273.16

Kerosene. a light, hydrocarbon fuel or solvent.

Killed Steel. Steel treated with a strong deoxidizing agent, such as aluminum or silicon, in order to reduce the oxygen content to a level so no reaction occurs between carbon and oxygen during solidification.

Kiln. A furnace in which ceramics are fired.

Kilo Calorie. This is the amount of heat (energy) necessary to raise the temperature of 1 kg of water 1□C. (Kilo calorie = kcal)

Kilo Calorie. This is the amount of heat (energy) necessary to raise the temperature of 1 kg of water 1□C. (Kilo calorie = kcal)

Kilo Grain. A unit of weight; 1000 grains.

Kilovolt Ampere. Product of the voltage times the current. Different from kilowatts because of inductive loads in an electrical system. Abbreviated: kVA kilo watts is equal to KVA times power factor. Kilowatt: 1000 watts. Abbreviated: kW.

Kilowatt. A metric unit of power equal to approximately 1.34 horsepower.

Kilowatt-Hour. A measure of electrical energy consumption. 1000 watts being consumed per hour. Abbreviated: kWh.

Kinetic Energy. The ability of an object to do work by virtue of its motion. (Water moving in a pipe has kinetic energy.) The energy terms that are usually used to describe the operation of a pump are "pressure" and "head". In classical mechanics, equal to one half of the body's mass times the square of its speed.

King Valve. Liquid receiver (refrigeration only) service valve.

Kirchoff's Second Law. The law stating that, at each instant of time the increase of voltage around a close loop in a network is equal to the algebraic sum of the voltage drop.

Knock. In a spark ignition engine, uneven burning of the fuel/air charge that causes violent, explosive combustion and an audible metallic hammering noise. Knock results from premature ignition of the last part of the charge to burn.

Knoop Hardness Test. An indentation hardness test using calibrated machines to force a rhombic-based pyramidal diamond indenter having specified edge angles, under specified conditions, into the surface of the test material and to measure the long diagonal after load removal.

Kraft Process. A wood-pulping process in which sodium sulfate is used in the caustic soda pulp-digestion liquor. Also called Kraft pulping or sulfate pulping.

kW Demand. The maximum rate of electric power usage required to operate a facility during a period of time, usually a month or billing period. Often called "demand".

kWh Consumption. The amount of electric energy used over a period of time; the number of kWh used per month. Often called "consumption".

L

Labyrinth Seal. A labyrinth seal consists of a number of rings 1 - 2 millimeters thick fixed to the shaft, tapered at the outer periphery to nearly knife-sharp with a clearance to the casing of a few hundreds of a millimeter. The rings are of brass or stainless steel, the sharp edge gives better sealing and rubs off easily without excessive heating in case of a slightly eccentric shaft. Some labyrinth seals are very simple, others are complicated.

Lag. A delay in the effect of a changed condition at one point in the system, on some other condition to which it is related. Also, the delay in action of the sensing element of a control, due to the time required for the sensing element to reach equilibrium with the property being controlled; i.e., temperature lag, flow lag, etc.

Laminar Flow. A non-turbulent flow regime in which the stream filaments glide along the pipe axially with essentially no transverse mixing.

Langelier Saturation Index. An index (SI) based upon the pH of saturation of calcium carbonate; used to determine the tendencies of a water supply toward corrosion or scaling. A positive index indicates scaling tendencies; a negative one means corrosion tendencies. (Langlier Index = pH - pH_s, where pH = actual pH of water and pH_s = pH at which water having the same alkalinity and calcium content is just saturated with calcium carbonate.

Lapping. A surface finishing operation used to achieve a fine polish and close tolerances.

Laser Soldering. Method of soldering in which the heat required to reflow a solder interconnection is provided by a laser (YAG or CO2); the solder joints are heated sequentially and cooled rapidly.

Laser. Acronym for light amplification by stimulated emission of radiation; a source of concentrated coherent light generated by stimulating electronic or molecular transitions to lower energy levels.

Latent Heat of Condensation. Amount of heat released (lost) by a pound of a substance to change its state from a vapor (gas) to a liquid.

Latent Heat of Fusion. The heat required to change 1.0 kg of a substance from the solid to the liquid state.

Latent Heat of Vaporization. The energy required to produce saturated vapor from saturated liquid at constant pressure per unit mass of fluid.

Latent Heat. Change of enthalpy during a change of state, usually expressed in Btu per lb. With pure substances, latent heat is absorbed or rejected at constant pressure.

Latent Heat. Heat energy absorbed in process of changing form of substance (melting, vaporization, fusion) without change in temperature or pressure.

Latent Heat. The thermal energy absorbed or released when a material experiences a phase change.

Lattice Parameter. The length of any side in a crystal structure's unit cell.

Lattice. The space arrangement of atoms in a crystal.

Law of Partial Pressure, Dalton's. Each constituent of a mixture of gases behaves thermodynamically as if it alone occupied the space. The sum of the individual pressures of the constituents equals the total pressure of the mixture.

Leaching. Process of liquid solder dissolving a metal coating.

Lead. A wire that connects two points in a circuit; it is usually self-supporting.

Leading Edge. Refers to the point where the steam enters the blade of an impulse turbine.

Leak Detector. Device or instrument such as a halide torch, an electronic sniffer; or soap solution used to detect leaks.

Leakage. 1. In water treatment, it refers to the passing of impure steam or boiler water through the drum internals.

2. In water treatment, the phenomenon in which some of the influent ions are not adsorbed and appear in the effluent when a solution is passed through an under regenerated exchange resin bed.

Lean NO_x Catalyst (LNC). Catalyst designed to reduce nitrogen oxides from diesel or spark-ignited engine exhaust gases under net oxidizing conditions, i.e., in the presence of excessive amount of oxygen.

Lever Rule. In a phase diagram, the relative proportions of the conjugate phases, at a stated value of temperature and pressure, or both, is such that a state of mechanical balance would obtain, if the corresponding weight of each phase were placed upon its composition point upon the tie-element (tie-line, tie-triangle, etc.) And the fulcrum were located at the gross composition point of the mixture.

Light Crude Oil. A crude oil of relatively high API gravity (usually 40°C degrees or higher).

Light Duty Vehicle (LDV). A passenger car or small truck.

Light Emitting Diode. A low current and voltage light used as an indicator on load management equipment. Abbreviated: LED.

Light Metal. One of the low-density metals, such as aluminum, magnesium, titanium, beryllium, or their alloys.

Lime. A common water treatment chemical.

Lime-Soda Softening. A process by which the calcium and magnesium salts, constituting the harness content of a water, are chemically precipitated and removed.

Line Voltage. In the control industry, the normal electric supply voltages, which are usually 120 or 240 volts.

Liquefaction. The change of state from a gas to a liquid. (The term liquefaction is usually used instead of condensation when referring to substances which are in a gaseous state at ordinary pressures and temperatures.)

Liquefied Natural Gas (LNG). Natural gas that has been refrigerated to cryonic temperatures where the gas condenses into a liquid.

Liquefied Petroleum Gas (LPG). Liquefied Petroleum Gas (LPG) is a mixture of low-boiling hydrocarbons that exists in a liquid state at ambient temperatures when under moderate pressures (less than 1.5 MPa or 200 psi). LPG is a by-product from the processing of natural gas and from petroleum refining. Major components of LPG are propane (min. 85% content in the U.S.), butane and propylene.

Liquid Absorbent. Chemical in liquid form which has the property to "take on" or absorb other fluids. Glycol is such a liquid and widely use in the petroleum chemical industry

Liquid Impingement. Material removal due to action of an impingement stream of a fluid.

Liquid Indicator. Device located in liquid line which provides a glass window through which liquid flow may be watched.

Liquid Line Charging Valve. The line used for charging from the high side of the refrigeration system.

Liquid Line. Tube which carries liquid refrigerant from the condenser or liquid receiver to the refrigerant control mechanism.

Liquid Nitrogen. Nitrogen in liquid form which is used as a low temperature refrigerant in expendable or chemical refrigerating systems.

Liquid Penetrant Inspection. A type of nondestructive inspection that locates discontinuities that are open to the surface of a metal by first allowing a penetrating dye or fluorescent liquid to infiltrate the discontinuity, removing the excess penetrant, and then applying a developing agent that causes the penetrant to seep back out of the discontinuity and register as an indication.

Liquid Receiver Service Valve. Two or three-way manual valve located at the outlet of the receiver and used for installation and service purposes. It is sometimes called the king valve.

Liquid Receiver. Cylinder (container) connected to condenser outlet for storage of liquid refrigerant in a system.

Liquid. Substance whose molecules move freely among themselves, but do not tend to separate like those of gases.

Liquidus. The line on the phase equilibrium diagram above which only liquids are stable and below which some solid is present; the lowest temperature at which a metal or alloy is completely liquid.

Liquid-Vapor Valve Refrigerant Cylinder. Dual hand valve on refrigerant cylinders, which is used to release either gas or liquid refrigerant from the cylinder.

Liquor. Solution used in absorption refrigeration.

Lithium Bromide. A chemical used in combination with water in absorption cooling systems.

Load (Air Conditioning). The amount of heat per unit time imposed on a refrigeration system or the required rate of heat removal.

Load Turbine (Gas). Is the turbine which is directly coupled to the load, which can only be the case with a multishaft gas turbine arrangement.

Load. The amount of heat per unit time imposed on a refrigeration system or the required rate of heat removal.

Locked Rotor Current. See "Inrush Current".

Longitudinal Seam. A riveted or welded seam along the longitudinal axis of a boiler shell or drum.

Low Emission Vehicle (LEV). A vehicle that is certified to meet the LEV emission standards set by the California Air Resources Board (CARB).

Low Side. The refrigerating system from the expansion point to the point where the refrigerant vapor is compressed; where the system is at or below evaporated pressure.

Low Temperature Cutout, Refrigeraton. A pressure or temperature actuated device with sensing element in the evaporator, which will shut the system down at its control setting to prevent freezing chilled water or to prevent coil frosting. Direct expansion equipment may not use this device.

Low Voltage. In the control industry, a power supply of 25 volts or less.

Low-Side Float Valve. Refrigerant control valve operated by level of liquid refrigerant in low-pressure side of system.

Low-Side Pressure Control. Device used to keep low side evaporating pressure from dropping below certain pressure.

Low-Side Pressure. Pressure in cooling side of refrigerating cycle.

Lubricant, Bonded Film. (See lubricant, bonded solid).

Lubricant, Bonded Solid. A solid lubricant dispersed in a continuous matrix of a binder or attached to a surface by an adhesive material.

Lubricant, Chlorinated. A lubricant containing a chlorine compound that reacts with a rubbing surface at elevated temperatures to protect it from sliding damage, (see extreme pressure lubricant).

Lubricant, Synthetic. A lubricant produced by synthesis rather than by extraction or refinement.

Lubricant. A substance - usually petroleum based, that is used to reduce friction between two moving parts.

Lubrication, Aerodynamic. (See gas lubrication).

Lubrication, Aerostatics. (See pressurized gas lubrication).

Lubrication, Bath. (See lubrication, flood).

Lubrication, Boundary. A condition of lubrication in which the friction and wear between two surfaces in relative motion are determined by the properties of the surfaces and by the properties of the lubricant other than bulk viscosity.

Lubrication, Drip Feed. A system of lubrication in which the lubricant is supplied to the bearing surface in the form of drops at regular intervals.

Lubrication, Dry Film. Lubrication that involves the application of a thin film of solid lubricant to the surface or surfaces to be lubricated.

Lubrication, Flood. A system of lubrication in which the lubricant is supplied in a continuos stream at low pressure and subsequently drains away.

Lubrication, Full Film. A type of lubrication wherein the solid surfaces are separated completely by an elstohydrodynamic fluid film.

Lubrication, Magneto Hydrodynamic. Hydrodynamic lubrication in which a significant force contribution arises from electromagnetic interaction.

Lubrication, Melt. Lubrication provided by steady melting of lubricating species. Also phase-change lubrication.

Lubrication, Mist. Lubrication by an oil mist produced by injecting oil into a gas stream.

Lubrication, Oil Fog. (See Mist Lubrication).

Lubrication, Oil Ring. A system of lubrication for horizontal shafts. A ring of larger diameter rotates with the shaft and collects oils from a container beneath.

Lubrication, Pad. A system of lubrication in which the lubricant is delivered to a bearing surface by a pad of felt or similar material.

Lubrication, Pressurized Gas. A system of lubrication in which a gaseous lubricant is supplied under sufficient external pressure to separate the opposing surfaces by a gas film.

Lubrication, Solid-Film. Lubrication by application of a solid lubricant.

Lubrication, Splash. A system of lubrication in which the lubricant is splashed onto the moving parts.

Lubrication, Thick Film. (also known as flood lubrication).

Lubrication, Thin Film. (also known as boundary lubrication).

Lubrication, Vapor-Phase. A type of lubrication in which one or more gaseous reactants are supplied to the vicinity of the surface to be lubricated and which subsequently react to form a lubricious deposit on that surface.

Lubrication, Waste. A system of lubrication in which the lubricant is delivered to a bearing surface by cloth waste or yarn.

Lubrication, Wick. A system of lubrication in which the lubricant is delivered to a bearing surface by means of a wick.

Lubricity. The ability of a lubricant to reduce wear and friction, other than by its pure viscous properties.

Luder Bands. Elongated surface markings or depressions in sheet metal caused by discontinuous yielding.

M

Macromolecule. A huge molecule made up of thousands of atoms.

Magnesium. A scale forming element found in some boiler feed water.

Magnetic Field Strength (*H*). The intensity of an externally applied magnetic field.

Magnetic Field. The region within which a body or current experiences magnetic force.

Magnetic Flux Density (*B*). The magnetic field produced in a substance by an external magnetic field.

Magnetic Flux. The rate of flow of magnetic energy across or through a surface.

Magnetic Iron Oxide (Fe_3O_2). Partially oxidized iron.

Magnetic Particle Inspection. A nondestructive method of inspection for determining the extent of surface cracks and similar imperfection in ferromagnetic materials.

Magnetic Pole. The area on a magnetized part at which the magnetic field leaves or enters the part. It is the point of maximum attraction in a magnet.

Magnetic Refrigeration. Where very low temperatures are obtained by using paramagnetic salts with magnets.

Magnetization. The total magnetic moment per unit volume of material; also, a measure of the contribution to the magnetic flux by some material within an *H* field.

Makeup Water. Water fed to a system to replace that which is lost - for example, water fed to a boiler to replace that lost as steam or condensate; water fed to a cooling tower to replace that lost by evaporation, drift, or other causes.

Malleability. The characteristic of metals that permits plastic deformation in compression without fracture.

Malleable Cast Iron. White cast iron that has been heat treated to convert the cementite into graphite clusters; a relatively ductile cast iron.

Manganese. A metallic element occasionally found in very small amounts as an impurity in well-water supplies.

Manifold, Service. Chamber equipped with gauges and manual valves, used by service technicians to service refrigerating systems.

Manifolding. A method of circulating the refrigerant through separate rows of tubes and mostly used with direct-expansion or dry evaporators.

Manometer. 1. A device to measure small to moderate pressure differentials. Device is general constructed from glass or plastic tubes filled with water, oil, alcohol or other suitable fluids.

2. An instrument for measuring pressures: especially a U-tube partially filled with a liquid, usually water, mercury, or a light oil, so constructed that the amount of displacement of the liquid indicates the pressure being exerted on the instrument.

Manual Frost Control. Manual control used to change operation of refrigerating system to produce defrosting conditions.

Martensite. A metastable iron phase supersaturated in carbon that is the product of a diffusionless (athermal) transformation from austenite.

Mass. The quantity of matter in a body as measured by the ratio of the force required to produce a given acceleration, to the acceleration.

Master (Central) Control. Control of all outlets from one point.

Matrix. The continuous phase in a composite or two-phase alloy microstructure in which a second phase is dispersed.

MCM. Thousand circular mill used to describe large wire sizes.

Mechanical Seals. A mechanical device used to control leakage from the stuffing box of a pump. Mechanical seals are usually made of two flat surfaces, one of which rotates on a shaft. The two flat surfaces are of such tolerances as to prevent the passage of water between them.

Mechanical Wear. Removal of material due to mechanical process under conditions of sliding, rolling, or repeated impact. Included are abrasive wear, fatigue wear and adhesive wear, but not the corrosive and thermal wear.

Melting Point. 1. For a given pressure, the temperature at which the solid and liquid phases of the substance are in equilibrium.

2. Temperature at atmospheric pressure at which a substance will melt.

Membrane. A barrier, usually thin, that permits the passage only of particles up to a certain size or of special nature.

Membrane. A thin sheet or layer.

Mer. The group of atoms that constitutes a polymer chain repeat unit.

Mercaptan. A compound chemically similar to alcohol, with sulfur replacing the oxygen in the chemical structure. Many mercaptans have an offensive odor and are used as deodorants in natural gas.

Metal. An opaque lustrous elemental chemical substance that is a good conductor of heat and electricity and, when polished, a good reflector of light; most elemental metals are malleable, ductile, and are generally denser than the other elemental substances; metals are structurally distinguished from nonmetals by their atomic bonding and electron availability; the electron band structure of metals is characterized by a partially filled valence band; the "free electrons" lost from the outer shells of metallic atoms are available to carry an electric current; the defining property of a metal is that it is an element with a positive thermal coefficient of resistivity, meaning the electrical resistivity of a metal continuously increases as temperature increases.

Metallic Bond. A primary interatomic bond involving the nondirectional sharing of nonlocalized valence electrons ("sea of electrons") which are mutually shared by all the atoms in the metallic solid.

Metallurgical Factor. The condition of the metal, such as inclusions, chemical segregation's, cold work and others, which have an impact upon the rate of electrochemical corrosion rates.

Metallurgy. The science and technology of metals and alloys.

Metastable. (1) nonequilibrium state of a material with respect to some transition, conversion, or reaction but stabilized kinetically either by rapid cooling or by some molecular characteristics; (2) possessing a state of pseudoequilibrium that has a free energy higher than that of the true equilibrium state.

Mica. A silicate material used with high pressure gauge glasses on boilers.

Micro Filtration. A membrane filtration process, which forces water through a porous barrier. Pores are usually between 0.1 to 20 m m, when used for water purification. For filtering purposes, pore sizes are .045 m m.

Microbar. A unit of pressure equal to 1 dyne/cm^2 (one millionth of the pressure of the atmosphere).

Micron. A unit of length, the thousandth part of 1 mm of the millionth of a meter.

Micron. One millionth of a meter (0.000001), and another term for micrometer (10^-6).

Microprocessor. A small computer used in load management to analyze energy demand and consumption such that loads are turned on and off according to a predetermined program.

Microscopy. The investigation of microstructural elements using some type of microscope, e.g. Scanning electron microscopy (SEM), light optical microscopy (LOM).

Microstructure. The microscopic assemblage of grains, grain boundaries, amorphous phases, pores, and inclusions, that make up a material.

Microvoid Coalescence (MVC). Occurs due to the nucleation of microvoids, followed by their growth and eventual coalescence; initiation is caused by particle cracking or interfacial failure between an inclusion or precipitate particle and the surrounding matrix.

Mild Steel. A low-carbon steel of ordinary production.

Mill Scale. A natural black iron oxide coating loosely adhering to the interior of new piping or tubes.

Miller Indices. A set of three integers (four for hexagonal) that designate crystallographic planes, as determined from reciprocals of fractional axial intercepts.

Mineral. A naturally occurring inorganic substance having specified chemical composition and crystalline structure.

Miscibility. The ability of two liquids, not mutually soluble, to mix.

Mix Bed Demineralizer. Having a mixture of cation and anion exchange resin in the same housing.

Mixed Dislocation. A dislocation that has both edge and screw components.

Mixture. A physical blend of two or more substances.

Modulating Control. A mode of automatic control in which the action of the final control element is proportional to the deviation, from set point, of the controlled medium.

Modulating Refrigeration Cycle. Refrigerating system of variable capacity.

Modulating. Type of device or control which tends to adjust by increments (minute changes) rather than by either "full on" or "full off" operation.

Modulus of Elasticity (*E*). 1. The measure of rigidity or stiffness of a material. 2. The ratio of stress to strain when deformation is totally elastic; also a measure of the stiffness of a material.

Modulus of Rupture. Breaking strength in a nonductile solid as measured by bending.

Moisture Indicator. Instrument used to measure moisture content of a refrigerant.

Molecule. A group of atoms that are bound together by primary interatomic bonds.

Mollier Diagram. An enthalpy-entropy or enthalpy-pressure chart showing the thermodynamic properties of a fluid.

Monomer. 1. A molecule consisting of a single mer. 2. A molecule, usually an organic compound, having the ability to join with a number of identical molecules to form a polymer.

Motoer Control Center. A single metal enclosed assembly containing a number of motor controllers and possibly other devices such as switches and control devices.

Mud Drum. A pressure chamber of a drum or header type located at the lower extremity of a water tube boiler and fitted with blowoff valve.

Mullion Heater. Electrical heating element mounted in the mullion. Used to keep mullion from sweating or frosting.

Multiple Stage Compressor. Compressor having two or more compressive steps. Discharge from each step is the intake pressure of the next in series.

Multiple System. Refrigerating mechanism in which several evaporators are connected to one condensing unit.

Multi-Shat Gas Turbine. A gas turbine having more than one shaft, like the dual shaft arrangement.

Multistage Compression Refrigeration System. Where the refrigerant is vaporized and condensed alternately and is compressed in the vapor phase.

Multistage Thermostat. A thermostat which controls auxiliary equipment for heating or cooling in response to a greater demand for heating or cooling.

N

Naphta. A volatile, flammable liquid hydrocarbon distilled from petroleum and used as a solvent or fuel.

National Ambient Air Quality Standards (NAAQS). Ambient standards for six pollutants including ozone, carbon monoxide, nitrogen dioxide, lead, particulate matter, and oxides of sulfur specifically regulated under the U.S. Clean Air Act of 1990. Urban areas are required to achieve attainment in regards tó ambient concentrations of these criteria pollutants.

Natural Circulation. The circulation of a boiler caused by differences in density. Also referred as thermal or thermally induced circulation.

Natural Convection. Movement of a fluid caused only by temperature differences (density changes).

Natural Draft Cooling Tower. Cools water by moving air at low velocities.

Natural Gas (NG). A mixture of hydrocarbon compounds and small quantities of various non hydrocarbon components existing in the gas phase or in solution with crude oil in natural underground reservoirs. The main component of natural gas is methane.

Natural Gas. A highly compressible, highly expandable mixture of hydrocarbons having a low specific gravity and occurring naturally in gaseous form. Besides hydrocarbon gases, natural gas may contain quantities of nitrogen, helium, carbon dioxide, hydrogen sulfide and water vapor.

Naturally Aspirated. A term used to describe a diesel engine in which air flows into the engine by means of atmospheric pressure only.

NC. Normally closed contacts of a relay.

Necking. Reduction of the cross-sectional area of a material in a localized area caused by uniaxial tension.

Negative Charge. The electrical potential which an atom acquires when it gains one or more electrons; a characteristic of an anion.

Net Positive Suction. The difference between total pressure and vapor pressure in a fluid flow, expressed in terms of equivalent height or "head".

Network Polymer. A polymer composed of trifunctional mer units that form three-dimensional molecules.

Neumann's Law. The symmetry of the physical properties of a crystal must include the symmetry of the point group of the crystal.

Neutral. The circuit conductor that is normally grounded or at zero voltage difference to the ground.

Neutralization Number. An ASTM number given to quenching oils that reflect the oil's tendency toward oxidation and sludging.

Neutralizer. A substance that will combine with an acid or alkali chemically, thus removing the acidity or alkalinity.

Neutralizing Amines. Are amines used to neutralize the acid generated by the dissolution of carbon dioxide.

Newton. The unit of force in the metric system. A newton is the force required to accelerate an object of 1 kilogram mass to a velocity of 1 meter per second in 1 second.

Nipple. A short, threaded tubular coupling, used for making connections between pipe joints.

Nitrogen Blanketing. Used with wet standby, where the space above the water level is filled with nitrogen at about 5 to 10 psig in order to keep the oxygen out.

Nitrogen Dioxide. Mildly poisonous gas (NO_2) often found in smog or automobile exhaust

Nitrogen Oxides (NO_x). Several air-polluting gases composed of nitrogen and oxygen which play an important role in the formation of photochemical smog. Nitrogen oxides are collectively referred to as "NO_x", where "x" represents a changing proportion of oxygen to nitrogen. Internal combustion engines are significant contributors to the worldwide nitrogen oxide emissions. For the purpose of emission regulations, NO_x is composed of colorless nitric oxide (NO), and the reddish-brown, very toxic and reactive nitrogen dioxide (NO_2). Other nitrogen oxides, such as nitrous oxide N_2O (the anesthetic "laughing gas"), are not regulated emissions.

NMHC. Non-Methane Hydrocarbons.

NMOG. Non-Methane Organic Gases.

NO. Normally open contacts of a relay.

Nobel Metal. A chemically inactive metal, such as gold.

Noble Metal. A metal with high resistance to chemical reaction, especially oxidation and solution by organic acids; sometimes called a precious metal.

No-Frost Freezer. Low-temperature refrigerator cabinet in which no frost or ice collects on freezer surfaces or materials stored in cabinet.

Noise. Any undesired sounds, usually of different frequencies, resulting in an objectionable or irritating sensation.

Nominal Size Tubing. Tubing measurement which has an inside diameter the same as iron pipe of the same stated size.

Nominal Size. A designated size that may be different from the actual size.

Nonattainment Area. A region that exceeds the U.S. National Ambient Air Quality Standards (NAAQS) for one or more criteria pollutants. Such regions, or areas, are required to seek modifications to their State Implementation Plans (SIPs), setting forth a reasonable timetable using means that are approved by the Environmental Protection Agency (EPA) to achieve attainment of NAAQS by a certain date. Under the Clean Air Act, if a nonattainment area fails to attain NAAQS, the EPA may superimpose a Federal Implementation Plan (FIP) with stricter requirements. Also, the EPA may impose fines, construction bans, or cut-offs in Federal grant revenues until the area achieves applicable NAAQS.

Non-Carbonated Hardness. Hardness in water caused by chlorides, sulfates, and nitrates of calcium and magnesium.

Non-Code Installation. Functional refrigerating system installed where there are no local, state, or national refrigeration codes in force.

Non-Condensables. Gaseous material not liquefied when associated water vapor is condensed in the same environment.

Noncrystalline. The solid state wherein there is no long-range atomic order; sometimes the terms *amorphous*, *glassy*, and *vitreous* are used synonymously.

Non-Destructive Inspection (NDI*)*. A procedure such as ultrasonic or radiographic inspection, for determining the quality of a material without permanently altering anything.

Nondestructive Testing. (NDT) a procedure for determining the quality or characteristics of a material, part, or assembly without permanently altering it or its properties; examples include ultrasonic and radiographic inspection.

Non-Ferrous Alloy. Alloy containing less then 50 % iron.

Non-Frosting Evaporator. Evaporator which never collects frost or ice on its surface. Uses only thermostatic expansion valves.

Non-Mechanical Refrigeration. Those that obtain the required high and low pressure by some method other than a mechanical compressor.

Non-Reactive Silica. Is a polymeric form of silica; thermally unstable which reverts to normal silica when heated. Difficult to detect, but may be present when boiler feedwater shows none, but boiler water reads silica.

Nonsteady-state Diffusion. The diffusion condition for which there is some net accumulation or depletion of diffusing species; the diffusion flux is dependent on time.

Normal Carryover. Refers to the carryover which occurs in any boiler operating under the best conditions.

Normal Charge. Thermal element charge which is part liquid and part gas under all operating conditions.

Normalizing. For ferrous alloys, austenitizing above the upper critical temperature, then cooling in air; the objective of this heat treatment is to enhance toughness by refining the grain size.

Normally Open (Or Normal Closed). The position of a valve, damper, relay contacts, or switch when external power or pressure is not being applied to the device. Valves and dampers usually are returned to a "normal" position by a spring.

NTU. Nephelometric turbidity unit. A light -interference analytical method to measure the turbidity of water.

Nucleate Boiling. The even boiling of water in which steam bubbles are formed within the boiler water gradually and are evenly distributed rather than being suddenly formed and erratically distributed.

Nucleation. The initial stage in a phase transformation; it is evidenced by the formation of small particles (nuclei) of the new phase which are capable of growing.

O

Occlusion. An absorption process by which one solid material adheres strongly to another, sometimes occurring by co precipitation.

Octahedral Position. The void space among close-packed, hard sphere atoms or ions for which there are six nearest neighbors; an octahedron (double pyramid) is circumscribed by the lines constructed from centers of adjacent spheres.

Octane Rating. A classification of gasoline according to its antiknock qualities. The higher the octane number or rating, the greater are the antiknock qualities of the gasoline.

Octyl Alcohol - Ethyl Hexanol. Additive in absorption machines to reduce surface tension in the absorber.

Offset. Term used to describe the difference between the set point and the actual operating or control point.

Ohm. The unit of electrical resistance equal to the resistance through which a current of 1 ampere will flow when there is potential difference of one volt across it.

Ohm's Law. The relationship between current and voltage in a circuit. It states that current is proportional to voltage and inversely proportional to resistance. Expressed algebraically, in DC circuits $I=E/R$; in AC circuits $I=E/Z$.

Oil Binding. Condition in which an oil layer on top of refrigerant liquid may prevent if from evaporating at its normal pressure temperature.

Oil Groove. A channel or channels in a bearing to improve oil flow through the bearing.

Oil Ring Lubrication. A system of lubrication for horizontal shafts. A ring of larger diameter rotates with the shaft and collects oils from a container below.

Oil Separator. Device used to remove oil from gaseous refrigerant or steam.

Oil, Mineral. A refined hydrocarbon oil without animal or vegetable additives.

Oil, Multi-Grade. An oil having relative little change in viscosity over a specified temperature range.

Oil, Neutral. A lubricating oil obtained by distillation, not treated with acid or with alkali.

Oil, Turbine. An oil used to lubricate bearings in a steam or gas turbine.

Oil. A liquid of vegetable, animal, mineral, or synthetic origin that feels slippery to the touch.

On-Board Diagnostics (OBD). A system on board of the vehicle that monitors emission control components and alerts the driver (e.g., by a dashboard light) if malfunctions or emission deterioration occurs. The OBD system involves a number of sensors and a data processor, which is typically integrated with the vehicle's electronic management system.

Once-Through Boiler. A steam generating unit usually operated above the critical pressure in which there is no re-circulation of the working fluid in any part of the unit.

On-Off Control. A two position action which allows operation at either maximum or minimum condition, or on or off, depending on the position of the controller.

Open Circuit. The absence of a direct connection between two points in an electrical network.

Open Cycle. A gas turbine arrangement, in were the exhaust gases from the turbine are exhausted to the atmosphere without any further treatment.

Open Frontal Area (OFA). In monolith catalyst substrates, the part of the total substrate cross-section area which is available for the flow of gas. The OFA is frequently expressed as a percentage of the total substrate cross-section and sometimes also called the substrate void fraction.

Open Re-Circulating Water System. A system, using continuously circulated water as a heat-transfer medium, in which the water is exposed at one point to the atmosphere for either discharge or absorption of heat.

Operating Point. The value of the controlled condition at which the controller actually operates. Also called control point.

Operating Pressure. Actual pressure at which the system works under normal conditions. This pressure may be positive or negative (vacuum).

Ordering. Positioning of host and substitution ions in an ordered, repetitious pattern rather than in a random arrangement.

Ore. A natural mineral mined and treated for extraction of its components.

Organic Growth. A substance resulting from the growth of biological organisms such as fungi, algae, and slime bacteria.

Organic Material. Contain carbon and usually hydrogen and are derived from living things.

Organic Oxygen Scavengers. These are organic compounds such as hydroquinone and ascorbate to remove dissolved oxygen from the boiler feedwater and condensate.

Orientation. Arrangements in space of the axes of a crystal lattice with respect to a coordinate system.

Orifice. Accurate size opening for controlling fluid flow.

Original Equipment Manufacturer (OEM). Manufacturers of equipment (such as engines, vehicles, etc.) that provide the original product design and materials for its assembly and manufacture. OEMs are directly responsible for manufacturing and modifying the products, making them commercially available, and providing the warranty.

Orsat Analyzer. A furnace atmosphere analysis device in which gases are absorbed selectively (volumetric basis) by passing them through a series of pre-selected solvents.

OSHA. Occupational Safety and Health Administration.

Osmosis. The passage of water through permeable membrane separating two solutions of different concentration; the water passes into the more concentrated solution.

Outgassing. Gaseous emission or de-aeration of a material.

Outside Air Opening (HVAC). Any opening used as an entry for air from outdoors.

Over Current Device. A device such as a fuse or a circuit breaker designed to protect a circuit against excessive current by opening the circuit.

Overflow Pipe. A pipe installed at a top of a tank to enable the liquid within to be discharged to another vessel when the tank is filled to capacity.

Overhead Cam. A camshaft used for operating both valves and unit injectors, located on top of or within the cylinder head. Such camshafts are driven by a multi-gear geartrain off the crankshaft. They simplify the design of the cylinder head and eliminate pushrods, allowing for much larger, open intake and exhaust ports and better breathing.

Overload. A condition of excess current; more cur rent flowing than the circuit was designed to carry.

Override. A manual or automatic action taken to by pass the normal operation of a device or system

Overspeed Trip. On steam turbines, a mechanism that provides absolute reliable overspeed protection by shutting off the steam supply.

Oxidation. A reaction in which there is an increase in valence resulting from a loss of electrons; often associated with the corrosion of metals, where the corroded metal forms an oxide; elevated temperatures increase the rate of oxidation.

Oxidation. The reaction of an element or substance with oxygen, e.g., iron is oxidized by reaction with oxygen to form rust (iron oxide).

Oxidizing Agent. Any substance such as oxygen, or chlorine, that will readily add (take on) electrons.

Oxidizing Atmosphere. A furnace atmosphere with an oversupply of oxygen that tends to oxidize materials placed in it.

Oxidizing Biocide. An agent, such as chlorine, which will kill bacteria by the chemical process of oxidation.

Oxyacetylene Welding. An oxyfuel gas welding process in which the fuel gas is acetylene.

Oxygen Concentration Cell. (See differential aeration cell).

Oxygen Concentration Cell. The surface area in contact with the media of higher oxygen concentration becomes the cathodic area, and the area with the lower oxygen concentration becomes the anode.

Oxygen Pitting. Damage caused due to the presence of oxygen in the feedwater. Damage results are small pit-like holes in the metal.

Oxygen Scavenger. A substance that will absorb oxygen by chemical reaction.

Oxygen Storage Capacity (OSC). A capacity of the catalyst washcoat to store oxygen at lean and to release it at rich condition. Typically provided by cerium oxide (ceria), which oscillates between an oxidized and reduced state, depending on the exhaust gas chemistry. The OSC is an important component of three-way catalysts, used to extend the catalyst window.

Oxygenated Fuel. Any fuel substance containing oxygen, such as ethanol, methanol, or biodiesel. Oxygenated fuel tends to give a more complete combustion of its carbon into carbon dioxide (CO_2), thereby reducing emissions of hydrocarbons and carbon monoxide. Oxygenated fuels may result in increased nitrogen oxides emissions.

Ozone (O_3). An oxygen molecule with three oxygen atoms. The stratosphere ozone layer, which is a concentration of ozone molecules located at 10 to 50 kilometers above sea level, is in a state of dynamic equilibrium. Oxygen molecules absorb ultraviolet (UV) light to form ozone which, in turn, decomposes back to oxygen. These processes absorb most of the ultraviolet light from the sun, shielding life from the harmful effects of UV radiation. Ozone is normally present at ground level in low concentrations. In cities where high level of air pollutants is present, the action of the sun's ultraviolet light can, through a complex series of reactions, produces harmful concentrations of the ground level ozone. The resulting air pollution is known as photochemical smog.

Ozone. Triatomic oxygen (0_3). Sometimes used in cold storage or air conditioning installations as an odor eliminator. Can be toxic in certain concentrations.

P

Package Units, (Refrigeration). Complete refrigerating system including compressor, condenser and evaporator located in the refrigerated space.

Packed Column. A tower filled with small objects, designed to obtain large surfaces per volume between rising vapors and a descending liquid.

Packing Gland. The metal part that compresses and holds packing in place in a stuffing box.

Packing. Material made usually of woven animal, plant, mineral or metal fiber and some type of lubricant, placed in rings around the shaft of a pump and used to control leakage from the stuffing box.

Packing. The fill in a confined space in a stripping vessel, ranging from simple shaped units such as rocks or slats to complex shapes that provide large surface area per unit volume.

Parallel Circuit. One where all the elements are connected across the voltage source. Therefore, the voltage on each element is the same but the current through each may be different.

Paramagnetism. Property of a material that, when placed in a magnetic field, is magnetized parallel to the field to an extent proportional to the field; this does not apply at very low temperatures or in extremely large magnetic fields.

Partial Pressures. Condition where two or more gases occupy a space and each one creates part of the total pressure.

Particulate Matter (PM). Particles formed by incomplete combustion of fuel. Compression ignition (diesel) engines generate significantly higher PM emissions than spark ignited engines. The particles are composed of elemental carbon, heavy hydrocarbons (SOF), and hydrated sulfuric acid ("sulfate particulates").

Pascal. The accepted metric unit of measurement or pressure and stress component in the measurement of viscosity. A Pascal is equal to a force of 1 Newton acting an area of 1 square meter. The symbol is *Pa*.

Pascal's Law. Pressure imposed upon a fluid is transmitted equally in all directions.

Passivating (Anodic) Inhibitors. A material capable of forming a protective oxide film on metal surfaces.

Passivation. The process of rendering a metal surface chemically inactive, either by electrochemical polarization or by contact with a passivating agent.

Passivator. 1. A type of corrosion inhibitor that appreciably changes the potential of a metal to a more noble (positive) value.

2. A type of corrosion inhibitor that appreciably changes the potential of a metal to a more positive value.

Passive-Active Cell. A corrosion cell in which the anode is a metal in that active state and the cathode is the same metal in the passive state.

Passivity. A condition on metal surfaces that inhibits electrochemical action between the metal and its environment, such as with boiler water.

Pathogenic Bacteria. Disease-causing bacteria.

Peak Demand. The greatest amount of kilowatts needed during a demand interval.

Peak Load Pricing. A pricing principle that charges more for purchases that contribute to the peak demand and, thereby, cause the expansion of productive capacity when the peak demand exceeds the peak capacity (less minimum excess capacity). In the electric power industry, this means charging more hr electricity bought on or near the seasonal peak of the utility or on or near the daily peak of the utility. The latter requires special meters; the former does not.

Peaking Load. Is the electrical load drawn on the system during high power usage. Usually on very hot or cold days or during the supper hour.

Peierls Stress. The stress required to move a dislocation.

Peltier Effect. When direct current is passed through two adjacent metals one junction will become cooler and the other will become warmer. This principle is the basis of thermoelectric refrigeration.

Perfect Gas. A hypothetical gas obeying the relation $pv = RT$.

Peritectic. An isothermal reversible reaction in metals where a liquid phase reacts with a solid phase to produce a single (and different) solid phase upon cooling.

Peritectoid. An isothermal reversible reaction where a solid phase reacts with a second solid phase to produce a single (and different) solid phase upon cooling.

Permanent Gases. Cryogenic refrigerants.

Permeability. (1) passage or diffusion of a gas, vapor, liquid, or solid through a material without physically or chemically affecting it; (2) term used to express various relationships between magnetic induction and magnetizing force; either absolute permeability or specific (relative) permeability.

Permeability. The ability of a body to pass fluid under pressure.

Petrographic Study of Boiler Scales. The systematic and descriptive study of rocks. This method also allows the identification of boiler scales, since scale can consist of a large variety of minerals

Petroleum Oil. (See mineral oils).

Petroleum. A generic term applied to oil and oil products in all forms. Examples include crude oil, lease condensate, unfinished oil, refined petroleum products, and natural gas plant liquids.

Pewter. Tin-base white metal containing antimony and copper; 1 to 8% Sb and 0.25 to 3% Cu.

pH of Saturation (pH_s). The pH at which a sample of water is saturated with a specific salt; for example, the pH of saturation of calcium carbonate is the pH of a saturated solution of calcium carbonate.

Ph. The negative logarithm of hydrogen-ion activity denoting the degree of acidity or basicity of a solution; at 25 degrees C 7.0 (on a scale of 0 to 14) is the neutral value, with decreasing values below 7.0 indicating increasing acidity, and increasing values above 7.0 indicating increasing basicity.

pH. The negative logarithm of the hydrogen-ion concentration of a solution; simply a measure of the relative acidity or alkalinity of a water solution. (pH 1 very acidic; pH 14, very basic; pH 7, neutral).

Phase Diagram. Graphical representation of the temperature and composition limits of phase fields in an alloy or ceramic system; it can be an equilibrium diagram, approximation to an equilibrium diagram, or a representation of metastable conditions or phases.

Phase Rule. This states that the maximum number of phases (*P*) that may coexist at equilibrium is two plus the number of components (C) in the mixture minus the number of degrees of freedom (*F*): $P + F = C + 2$.

Phase Transformation. Changes that can occur within a given material system; how one or more phases in an alloy change into a new phase or mixture of phases; transformation occurs because the initial state of the alloy is unstable relative to the final state; at constant temperature and pressure the relative stability of a system is determined by its Gibbs free energy.

Phase. 1. A portion of a material system whose properties and composition are homogeneous and which is physically distinct from other parts of the system.

2. Part of an AC voltage cycle. Residential electrical service is 2-phase; commercial facilities are usually 3-phase AC voltage.

Phial -Term sometimes used to denote the sensing element on a thermostatic expansion valve.

Phosphate Treatment. An internal boiler water treatment method to reduce calcium in the boiler with low hardness feedwater.

Phosphate. An ion, compound, or salt containing phosphorus and oxygen, such as sodium phosphate (Na_3P0_4).

Photolithography. Science of replicating complex circuitry onto the surface of a specimen.

Physical Stability. The quality which an ion-exchange resin must possess to resist changes that might be caused by attrition, high temperatures, and other physical conditions.

Physical Water Treatment. Refers to the treatment of removing dissolved gases from the boiler feedwater, using steam.

Pickling. The chemical removal of surface oxides and other contaminants from a material by immersion in an aqueous acid solution; sulfuric and hydrochloric acids are common pickling solutions.

Piezoelectricity. Elastic strain caused when an electrical current is applied to a piezoelectric material and conversely, an electric current produced when pressure is applied to a piezoelectric material; these materials exhibit the Perovskite crystal structure.

Pig Iron. High-carbon iron made by reduction of iron ore in the blast furnace.

Pig. A metal casting used in remelting.

Pilot Duty Relay. A relay used for switching loads such as another relay or solenoid valve coils. The pilot duty relay contacts are located in a second control circuit. Pilot duty relays are rated in volt-amperes (VA).

Pipe Scale. Rust or mill scale found on the interior of water pipe.

Pit. Corrosion localized in a small spot.

Pitote Tube. An open ended tube arrangement to face against the current of a stream of fluid; used in measuring the velocity head of a flowing medium.

PITS. Petroleum Industry Training Service. (Canada)

Pitting. Localized corrosion of a metal characterized by small blisters under which holes have perforated the metal.

Plane Strain. Thin plate and ?z = 0; 2D strain / 3D stress. (see Griffith Crack Theory)

Plane Stress. Thick sections; 2D stress / 3D strain. (see Griffith Crack Theory)

Plastic Deformation. The permanent (inelastic) distortion of a material under an applied stress that strains the material beyond its elastic limit; the ability of a material to be permanently deformed without fracture.

Plasticity. The ability of a substance to be deformed without rupturing.

Plating. Forming an adherent layer of metal on an object; often used as a shop term for electroplating.

PLENUM Chamber. An air compartment connected to one or more distributing ducts.

Plenum Chamber. Chamber or container for moving air or other gas under a slight positive pressure.

Pneumatic. Operated by air pressure.

Pneumatic-Electric (PE) Switches. Device that operates an electric switch from a change of air pressure.

pOH. An expression of the alkalinity of a solution; the negative logarithm of the hydroxyl-ion concentration.

Point, Critical. Of a substance, state point at which liquid and vapor have identical properties; critical temperature, critical pressure, and critical volume are the terms given to the temperature, pressure, and volume at the critical point. Above the critical temperature or

Poisson's Ratio. The ratio of the transverse contracting strain to the longitudinal elongational strain when a tensile stress is applied to a material.

Polarity. The direction of current flow in a DC circuit. By convention, current flows from plus to minus. Electron flow is actually in the opposite direction.

Polarization. Displacement of the centers of positive and negative charge.

Polarize. In corrosion, to develop a barrier on the anodic or cathodic surface, disrupting the corrosion process.

Pole. An electrical connection point. In a panel, the point of connection. On a device, the terminal that connects to the power.

Polycyclic Organic Matter (POM). A class of air toxics defined in the US Clean Air Act as compounds with more than one benzene ring and a boiling point of 100□C and higher. Includes practically all of diesel PAH material.

Polyelectrolyte. A water-soluble polymer that is highly charged with cationic or anionic sites. Used as an aid with coagulants.

Polyimide. Thermosetting ring chain polymer characterized by -NH group; it's increasingly used as dielectrics in high performance circuits.

Polymer. A chemical compound formed by the adjoining of many molecules of the same characteristics; for example, polyvinyl chloride is a polymer of the molecule vinyl chloride.

Polymeric Dispersant. A water-soluble polymer that acts as a suspending agent; that is, it promotes suspension of particles in water so that they resist settling.

Polymorphism. Different crystal structures at different temperatures or pressures for a single compound.

Polynuclear Aromatic Hydrocarbons (PAH). Aromatic hydrocarbons with two or more (up to five or six) benzene rings joined in various, more or less clustered forms.

Polyphosphate. A chemical compound formed by the adjoining of phosphate ions, hence a polymer of phosphate.

Polytrophic Process. A non-adiabatic reversible process characterized by the equation of path, pv = constant.

Porosity. Fine holes, voids, interstitials, or open spaces between grains or trapped in grains of a material's microstructure.

Porous Vase. A vase having a porous construction, which was used to cool its content.

Positive Charge. The electrical potential acquired by an atom which has lost one or more electrons; a characteristic of a cation.

Potential Energy. The energy, that a body has by virtue of its position or state, enabling it to do work. (Water in a reservoir has potential energy by virtue of its elevation above some other point.)

Potential Transformer. A voltage transformer. The voltage supplied to a primary coil induces a voltage in a secondary coil according, to the ratio of the wire windings in each of the coils.

Potentiometer. An electromechanical device having a terminal connected to each and to the resistive element, and a third terminal connected to the wiper contact. The electrical input is divided as the contact moves over the element, thus making it possible to mechanically change the resistance.

Pour Point. 1. A measure of the ability of a diesel fuel to operate under cold weather conditions. Defined as the temperature at which the amount of wax out of solution is sufficient to gel the fuel when tested under standard conditions (ASTM D97).

2. The lowest temperature at which a lubricant can be observed to flow under specified conditions. For oil, the pour point is a temperature -15□C above the temperature at which the oil is solid.

Pour-Point Depressant. An additive that lowers the pour point of a lubricant.

Power (P). Expressed in watts (W) or kilowatts (kW), and is equal to:

Power Element. Sensitive element of a temperature-operated control.

Power Factor (pf). A quantity that relates the volt amperes of an AC circuit to the wattage (power = volt-amperes x power factor). Power factor also is the ratio of the circuit resistance (R) to the impedance (Z) expressed as a decimal between zero and one (pf = R/Z). When the power factor equals one, all consumed power produces useful work.

Power Factor Charge. A utility charge for "poor" power factor. It is more expensive to provide power to a facility with a poor power factor (usually less than 0.8).

Power Factor Correction. Installing capacitors on the utility service's supply line to improve the power factor of the building.

ppb. equals 0.001 ppm. (Parts per billion)

Precipitate. An insoluble reaction product; in an aqueous chemical reaction, usually a crystalline compound that grows in size to become settable.

Precipitating (Cathodic) Inhibitors. The are chemicals which form insoluble precipitates that are able to coat and protect surfaces. They are less durable then the passivating type inhibitors.

Precipitation Hardening. Increase the hardness of a supersaturated solid solution by heat treating it to cause a second phase to precipitate out; coherency of the precipitate/matrix interface and how well the two lattices match up greatly influence the effect of precipitate.

Precipitation Heat Treatment. Artificial aging of metals in which a constituent precipitates from a supersaturated solid solution.

Precipitation. 1. A process whereby salts drop or come out of a water solution. 2. In metals, the separation of a new phase from solid or liquid solution, usually with changing conditions of temperature, pressure, or both.

Precision Phosphate Treatment. A phosphate treatment based upon maintaining 2 to 4 ppm of phosphate and 15 to 50 ppm of hydrate alkalinity in the boiler.

Precombustion Chamber. A small, auxiliary combustion chamber connected by a narrow orifice with the main chamber. Fuel is injected into the prechamber and ignites there, causing hot gases to expand into the main chamber (cylinder).

Predicting Method. A method for determining when load shedding should occur. A formula is used to arrive at a preset kilowatt limit. Then the actual amount of energy accumulated during the utility's demand intervals is measured. A projection is made of the actual rate of energy usage during the rest of the interval. If the predicted value exceeds the preset limit, loads will be shed.

Preignition. A condition in an internal combustion engine characterized by a knocking sound and caused by the fuel-air mixture having been ignited to soon because of an abnormal condition.

Pressure Drop. Pressure loss in fluid pressure, as from one end of a duct or pipe to the other, due to friction, dynamic losses, and changes in velocity pressure.

Pressure Gauge. Instrument for measuring the pressure exerted by the contents on its container.

Pressure Head. The height to which liquid can be raised by a given pressure (sometimes referred to as pump head).

Pressure Motor Control. Device which opens and closes an electrical circuit as pressures change.

Pressure Operated Thermometer. Temperature indicator which is controlled by bellows, a capillary or remote sensitive bulb.

Pressure Regulator, Suction (Refrigeration Compressors). An automatic valve designed to limit the suction pressure to prevent motor overload.

Pressure Switch. Switch operated by a rise or drop in pressure.

Pressure Velocity. In moving fluid, the pressure capable of causing an equivalent velocity, if applied to move the same fluid through an orifice such that all pressure energy expended is converted into kinetic energy.

Pressure, Absolute. Pressure referred to that of a perfect vacuum. It is the sum of gauge pressure and atmospheric pressure.

Pressure, Atmospheric. It is the pressure indicated by a barometer. Standard atmosphere is the pressure equivalent.

Pressure, Critical. Vapor pressure corresponding to the substance's critical state at which the liquid and vapor have identical properties.

Pressure, Gage. Gauge: Pressure above atmospheric.

Pressure, Hydrostatic. The normal force per unit area that would be exerted by a moving fluid on an infinitesimally small body immersed in it if the body were carried along with the fluid.

Pressure, Operating. Pressure at which a system is operating.

Pressure, Partial. Portion of total gas pressure of a mixture attributable to one component.

Pressure, Saturation- The saturation pressure for a pure substance for any given temperature is that pressure at which vapor and liquid, or vapor and solid, can coexist in stable equilibrium.

Pressure, Static. The normal force per unit area that would be exerted by a moving fluid on a small body immersed in it if the body were carried along with the fluid. Practically, it is the normal force per unit are at a small hole in a wall of the duct through which the fluid flows (piezometer) or on the surface of a stationary tube at a point where the disturbances, created by inserting the tube, cancel. It is supposed that the thermodynamic properties of a moving fluid depend on static pressure in exactly the same manner as those of the same fluid at rest depend upon its uniform hydrostatic pressure

Pressure, Total. In the theory of the flow of fluids, the sum of the static pressure and the velocity pressure at the point of measurement. Also called dynamic pressure.

Pressure, Vapor. The partial pressure exerted by the water vapor contained in air.

Pressure. The normal force exerted by a homogeneous liquid or gas, per unit of area, on the wall of its container.

Pressure-Heat Diagram. Graph of refrigerant pressure, heat and temperature properties. (Mollier's diagram.)

Pressurestat. A control which reacts to pressure changes in the evaporator.

Pretreatment. Also referred to as external treatment, consisting of treating the raw make-up water, and include removing dissolved oxygen, carbon dioxide, suspended solids, hardness, alkalinity, silica, dissolved solids, etc..

Primary Air. The initial air stream discharged by an air outlet (the air being supplied by a fan or supply duct) prior to any entrainment of the ambient air or for he purpose of combustion.

Primary Constituents. Refers to the category of dissolved solids present in water at a level of more then 5 mg/L.

Primary Control. A device which directly or indirectly controls the control agent in response to needs indicated by the controller. Typically a motor, valve, relay, etc.

Primary Element. The portion of the controller which first uses energy derived from the controlled medium to produce a condition representing the value of the controlled variable; for example, a thermostat bimetal.

Primary System. A closed water system in which the water is circulated directly through a chiller for cooling or heat exchanger or boiler for heating.

Prime Mover. An engine or other device by which natural source of energy is converted into mechanical energy.

Priming. The phenomenon in which the level of water in a boiler is changed by bouncing rapidly. The result is, that boiler water will enter the steam flow.

Process Hot Water. Hot water needed for manufacturing processes over and above the "domestic hot water" that is for the personal use of industrial workers.

Propane (C_3H_8). A normally gaseous straight-chain hydrocarbon. Propane is a colorless paraffinic gas that boils at a temperature of -42□C. It is extracted from natural gas or refinery gas streams.

Propane. A paraffin hydrocarbon (C_3H_8) that is a gas at ordinary atmospheric conditions but easily liquefied under pressure.

Proportional Band. The range of values of a proportional positioning controller through which the controlled variable must pass to move the final control element through its full operating range. Commonly used equivalents are "throttling range" and "modulating range".

PSI. Symbol or initials used to indicate pressure measured in pounds per square inch.

PSIA. Symbol or initials used to indicate pressure measured in pounds per square inch absolute. Absolute pressure equals gauge pressure plus atmospheric pressure. The "*A*" indicates that the gauge pressure is reading in absolute.

PSIG. Symbol or initials used to indicate pressure in pounds per square inch gauge. The "*G*" indicates that is gauge pressure and not absolute pressure.

Psychrometer or Wet Bulb Hygrometer – An instrument for measuring the relative humidity of atmospheric air.

Psychrometer. An instrument for ascertaining the humidity or hygrometric state of the atmosphere.

Psychrometric Chart – A chart that shows relationship between the temperature, pressure and moisture content of the air.

Psychrometric Measurement. Measurement of temperature pressure and humidity using a psychometric chart.

Psychrometry. The study of air-vapor mixtures.

Pull Down. An expression indicating action of removing refrigerant from all or part of a refrigeration system.

Pulsation Dampener. Any gas or liquid charged, chambered device that minimizes periodic increases and decreases in pressure.

Pulse-Echo Techniques. Corrosion detecting processes which, by recording the action of ultrasonic waves artificially introduced into production structures, can determine metal thickness and detect flaws.

Pump Down. The act of using a compressor or a pump to reduce the pressure in a container or a system.

Pump, Positive Displacement. A pump that moves a measured quantity of fluid with each stroke of a piston or each revolution of vanes or gears.

Pump, Rotary. A pump that moves fluid by positive displacement, using a system of rotating vanes, gears, or lobes.

Pump. A device that increases the pressure on a fluid or raises it to a higher level.

Pump-Discharge Pressure. The point of highest pressure in a re-circulating water system which is at the discharge side of the re-circulating pump.

Pump-Line-Nozzle Fuel System. A fuel system using a single injection pump driven off the geartrain on the front of the engine that also drives the camshaft. The central injection pump feeds a separate injection nozzle located in the cylinder head above each cylinder. Lines which must be of exactly equal length link each pump plunger with the associated nozzle. Each nozzle incorporates a needle valve and the orifices which actually handle atomization.

Purge Test. A test used to determine if fuel vapors are properly drawn from the evaporative canister and the fuel tank into the engine for combustion. If the purge system is not working properly, the evaporative canister can become saturated and vent hydrocarbons into the atmosphere.

Purging Valves. Devices used to remove non-condensable gases from the refrigeration system.

Purging. Releasing compressed gas to atmosphere through some part or parts for the purpose of removing contaminants from that part or parts.

Pyrometer. A device for measuring temperatures above the range of liquid thermometers.

Q

Quality. Weight fraction of the vapor in a vapor-liquid mixture.

Quench. Rapidly cool a material; typically done to retain a structure at room temperature that otherwise is only stable at high temperature.

Quicklime. Unslaked lime (calcium oxide).

R

Raceway. Any support system, open or closed, for carrying electric wires.

Radiant Heating. Heating system in which warm or hot surfaces are used to radiate heat into the space to be conditioned.

Radiation. Transfer of heat by electromagnetic waves.

Radioactivity. Spontaneous decay of some isotopes in nuclei.

Radiography. Nondestructive method of internal examination in which metal objects are exposed to a beam of X-ray or gamma radiation; differences in thickness, density, or absorption caused by internal defects or inclusions are apparent in the shadow image produced on a fluorescent screen or photographic film placed behind the object.

Ram Air. Air forced through the condenser due to the rapid movement of steam from the turbine exit.

Rankine Scale. Name given the absolute (Fahrenheit) scale. Zero (0) R on this scale is -460□F.

Rankine Temperature. Degrees Fahrenheit plus 491.60.

Rapture Membrane. A metal membrane (pressure relief device) designed to rapture at a predetermined pressure.

Raw Makeup Water. Untreated water fed to a system to replace that lost.

Raw Water. With water treatment it means untreated feedwater or water in its natural state, prior to any treatment.

Reagent. A substance, chemical, or solution used in laboratory to detect, measure, or react with other substances, chemicals or solutions.

Re-Carbonation. A process using carbon dioxide gas to stabilize and thereby prevent precipitation of calcium carbonate from cold process lime or lime-soda softened water.

Receiver Dehydrator. Small tank which serves as liquid refrigerants reservoir and which also contains a desiccant so remove moisture. Used on most automobile air conditioning installations.

Receiver Heating Element. Electrical resistance mounted in or around liquid receiver. It is used to maintain head pressures when ambient temperature is low.

Receiver. An auxiliary storage receptacle for fluids.

Receiver-Drier. Cylinder (container) in a refrigerating system for storing liquid refrigerant and which also holds a quantity of desiccant.

Reciprocal Lattice. A group of points arranged about a center in such a way that the line joining each point of the center is perpendicular to a family of planes in the crystal, and the length of this line is inversely proportional to their interplanar distance.

Reciprocal. A format defined by 1 divided by the complex number.

Reciprocating Compressor. Compressor which uses a piston and cylinder mechanism to provide pumping action.

Recording Thermometer. Temperature measuring instrument which has a pen marking a moving chart.

Rectifier. A device used to convert AC current into DC current..

Recuperator. Equipment for transferring heat from gaseous products of combustion to incoming air or fuel. The incoming material passes through pipes surrounded by a chamber through which the outgoing gases pass.

Red Water. Water that has a red, cloudy appearance caused by suspended red iron oxide.

Redox Potential. A process designed to determine if a corrosion process will occur.

Reducing Elbow. A fitting that makes an angle between two joints of pipe and that decreases in diameter from one end to the other.

Reducing Flange. A flange fitting used to join pipes of different diameters.

Reducing Nipple. A pipe fitting that is threaded on both ends and decreases in diameter from one end to the other.

Reducing Tee. A "T"shaped pipe fitting with openings of two different sizes. The relationship:

Reduction. Reduction is the addition of hydrogen, removal of oxygen, or the addition of electrons to an element or compound.

Reed Valve. Thin, flat, tempered steel plate fastened at one end.

Reflow Soldering. Process of joining metallic surfaces through the mass heating of solder/solder paste to form solder fillets at metallized areas; it creates a mechanical and electrical connection between components and a PCB.

Reformulated Gasoline (RFG). Gasoline whose composition has been changed, from that of gasolines sold in 1990, to (1) include oxygenates, (2) reduce the content of olefins, aromatics and volatile components, and (3) reduce the content of heavy hydrocarbons to meet performance specifications for ozone-forming tendency and for release of toxic substances (benzene, formaldehyde, acetaldehyde, 1,3-butadiene, and polycyclic aromatic hydrocarbons) into the air from both evaporation and tailpipe emissions.

Refractory. A heat-resistant material.

Refrigerant Charge. Quantity of refrigerant in a system.

Refrigerant Control. Device which meters flow of refrigerant between two areas of a refrigerating system. It also maintains pressure difference between high-pressure and low-pressure side of the mechanical refrigerating system while unit is running.

Refrigerant. 1. The fluid used for heat transfer in a refrigerating system, which absorbs heat at a low temperature and a low pressure of the fluid and rejects heat at a higher temperature and a higher pressure of the fluid, usually involving changes of state of the fluid. 2. The working fluid used in refrigerators.

Refrigerating Capacity. The ability of a system to remove heat as compared with the cooling effect produced by melting of ice.

Refrigerating Effect. The amount of heat transferred by one kg of refrigerant as it circulates in the refrigeration system.

Refrigeration Oil. Specially prepared oil used in refrigerator mechanism which circulates, to some extent, with refrigerant.

Refrigeration. Controlled transfer of heat from a lower temperature to a higher temperature region.

Refrigerator. A device to transfer heat from a low temperature to a high temperature medium.

Regenerant. The solution used to restore the activity of an ion exchanger. Acids are employed to restore a cation exchanger to its hydrogen form; brine solutions may be used to convert the cation exchange to the sodium form. The anion exchanger may be rejuvenated by treatment with an alkaline solution.

Regeneration. Restoration of the activity of an ion exchanger by replacing the ions adsorbed from the treated solution by ions that were adsorbed initially on the resin.

Regenerative Cycle. Is a gas turbine cycle employing a heat exchanger to recover some of the heat before discharging the gases from the gas turbine to the atmosphere, to recover some of the from

Regenerative Gas Turbine. Referring to a gas turbine employing heat exchanger between the compressor and the combustor for the purpose of recovering heat.

Register. A grille equipped with an integral damper or control valve.

Rejuvenation. (See regeneration).

Relative Humidity (RH). The ratio of water vapor in the air as compared to the maximum amount of water vapor that may be contained.

Relative Humidity. The amount of moisture the air holds relative to the maximum moisture the air can hold at the same temperature.

Relative Humidity. The ratio, expressed as a percentage, of the amount of water vapor present in a given volume of air at a given temperature to the amount required to saturate the air at that temperature.

Relay, Thermal. A switching relay in which a small heater warms a bimetal element which bends to provide the switching force.

Relay. An electromechanical switch that opens or closes contacts in response to some controlled action. Relay contacts can be normally open (NO) and/ or normally closed (NC). Relays may be electric, pneumatic, or a combination of both.

Relief Valve. Safety device on a sealed system. It opens to release fluids before dangerous pressure is reached. Also called pressure relief valve.

Remote System. Refrigerating system in which condensing unit is away from space to be cooled.

Remote Temperature Set Point. Ability to set a temperature control point for a space from outside the space. Often used in public areas.

Renewable Energy. Energy obtained from sources that are essentially inexhaustible, unlike fossil fuels. It includes conventional hydro-electric, wood, bio-feedstocks, waste, geothermal, wind, photovoltaic, and solar thermal energy.

Reset Ratio. The ratio of change in outdoor temperature to the change in control point temperature. For example, a 2:1 reset ratio means that the control point will increase 1 degree for every 2 degrees change in outdoor temperature.

Reset. A process of automatically adjusting the control point of a given controller to compensate for changes in outdoor temperature. The hot deck control point is normally reset upward as the outdoor temperature drops. The cold deck control point is normally reset downward as the outdoor temperature increases.

Residual stress. Internal stress in a material often resulting from thermal or mechanical straining.

Residual. Means small amount of, like oxygen, sulfite, acid., etc..

Resilience. The tendency of a material to return to its original shape after the removal of a stress that has produced elastic strain.

Resin Flux. A resin and small amounts of organic activators in an organic solvent.

Resin. An organic polymer that crosslinks to form a thermosetting plastic when mixed with a curing agent.

Resistance Welding. Type of welding process in which the work pieces are heated by the passage of an electric current through the contact; this includes spot welding, seam or line welding, and percussion welding; flash and butt welding are sometimes considered as resistance welding processes.

Resistance, Thermal. The reciprocal of thermal conductance.

Resistance. The opposition which limits the amount of current that can be produced by an applied voltage in an electrical circuit, measured in ohms.

Resistive Loads. Electrical loads whose power factor is one. Usually contain heating elements.

Resonant Vibration. Everything has a natural frequency. This frequency is effected by two properties: Mass and Stiffness. This "natural frequency" is the cause of many vibration problems in HVAC equipment. If you strike an object (say a tuning fork or a bell) it will continue to vibrate at its natural frequency until damping extinguishes the vibration.

Respirable Combustible Dust (RCD). A method of measuring ambient DPM exposures using a combustion process. Used in underground mines in Canada.

Restrictor. A device for producing a deliberate pressure drop or resistance in a line by reducing the cross-sectional flow area.

Return Air. Air returned from conditioned or refrigerated space.

Return-Steam Condensate. That steam produced by a boiler which returns to the boiler after it has condensed.

Reverse Cycle Defrost. Method of heating evaporator for defrosting. Valves move hot gas from compressor into evaporator.

Reverse Cycle Refrigeration. A refrigeration system which uses reject heat to produce warmth.

Reverse Deionization. The use of an anion-exchange unit and a cation-exchange unit-in that order-to remove all ions from solution.

Reverse Osmosis. A process that reverses (by the application of pressure) the flow of water in the natural process of osmosis so that it passes from the more concentrated to the more dilute solution.

Reversible Process. A process by which a fluid is made to undergo a change of state and by traversing the path in exactly the reverse of the original path is returned to its original state, and all associated systems are similarly returned to their original state.

Reversing Valve. Device used to reverse direction of the refrigerant flow depending upon whether heating or cooling is desired.

Rheology. Study of flow characteristics.

Rinse. The operation which follows regeneration; a flushing out of excess regenerant solution.

Rosin Flux. Rosin in an organic solvent or rosin as a paste with activators.

Rosin. A hard, natural resin, consisting of abietic acid and pimaric acids and their isomers, some fatty acids, and terepene hydrocarbons; the resin is extracted from pine trees and subsequently refined.

Rotary Compressor. Mechanism which pumps fluid by using rotating motion.

Rotary Injection Pump. A lower-cost injection pump used with pump-line-nozzle systems. The pump has a central plunger system (usually consisting of two opposing plungers) that provides fuel to every cylinder during the required injection period. A plate located near the top of the pump rotates, opening an appropriate orifice at the right time for distribution to each cylinder's injection nozzle through a separate line. It is usually used with automotive or agricultural engines that have lower performance and durability requirements than the heavy-duty truck diesels.

Running Current. The current that flows through a load after inrush current. Usually called "full load current".

Running Time. Amount of time a condensing unit is run per hour or per 24 hours.

Rust. A visible corrosion product consisting of hydrated oxides of iron. Applied only to ferrous alloys.

Ryznar Stability Index. An index based on calcium carbonate pH of saturation vs. actual pH to determine scaling or corrosion tendencies of a water (R.I. = 2 pHs).

S

Sacrificial Anodes. Coupling of a more active metal to a structure resulting in a galvanic current flow through the corroding electrolyte.

Sacrificial Protection. Reduction of corrosion of a metal in an electrolyte by galvanically coupling it to a more anodic metal. A form of cathodic protection.

Saddle Valve (Tap-A-Line). Valve body shaped so it may be silver brazed or clamped onto a refrigerant tubing surface.

Saddle Valve (Tap-A-Line). Valve body shaped so it may be silver brazed or clamped onto a refrigerant tubing surface.

Safety Can. Approved container of not more than 5 gallon capacity. It has a spring closing lid and spout cover. It is designed to relieve internal pressure safely when exposed to fire.

Safety Control. Device to stop refrigerating unit if unsafe pressure and/or temperatures and/or dangerous conditions are reached.

Safety Plug. Device which will release the contents of a container before rupture pressures are reached.

Salinity. A measure of the concentration of dissolved mineral substances in water.

Salt Spitting. The ability of an anion exchanger to convert a salt solution to caustic; the ability of a cation exchanger to convert a salt solution to acid.

Saponifier. An alkaline chemical added to water to improve its ability to dissolve rosin flux residues.

Saturated Air. When the air cannot hold any more moisture.

Saturated Liquid. A liquid which is at saturation pressure and saturation temperature; in other words, a liquid which is at its boiling point for any given pressure.

Saturated Vapor. A vapor which is at saturation pressure and saturation temperature. A saturated vapor cannot be superheated as long as it is in contact with the liquid from which it is being generated.

Saturation Index. The relation of calcium carbonate to the pH, alkalinity, and hardness of a water to determine its scale forming tendency.

Saturation Pressure. The point, where at a given temperature a pure substance starts to boil.

Saturation Temperature. The point, where at a given pressure a pure substance starts to boil.

Saturation. Condition existing when substance contains all of another substance it can hold for that temperature and pressure.

Saybolt Universal Viscosity. A commercial measure of viscosity expressed as a time in seconds required for 60 mL of a fluid to flow through the orifice of the standard Saybolt universal viscometer at a given temperature under specific conditions; used for the lighter petroleum products and lubrication oils.

Scale Causing Elements. Calcium and magnesium elements forming scale.

Scale Removal. Waterside, removal of scale using either the mechanical, the water treatment or the acid cleaning process.

Scale. Surface oxidation, consisting of partially adherent layers of corrosion products, left on metals by heating or casting in air or in other oxidizing atmosphere. Also a deposit on a heat-transfer surface resulting from precipitation of salts present in water in contact with that surface, forming a hard, dense material.

SCAQMD. South Coast Air Quality Management District.

SCC. Stress corrosion cracking.

Screens. Equipment designed to prevent larger objects to enter water treatment system.

Screw Pump. Compressor constructed of two mated revolving screws.

Scrubber. An apparatus for the removal of solids from gases by entrainment in water.

Seal, Magnetic. A seal that uses magnetic material, instead of springs or bellows, to provide the closing force.

Seal, Rotary. A mechanical seal that rotates with a shaft and is used with a stationary mating ring.

Second Law of Thermodynamics. Heat will flow only from material at higher temperature to material at lower temperature.

Secondary Refrigerant. Chilled liquid like water, which is circulated to distance units where air is to be cooled in individual rooms.

Secondary System. A re-circulating water system that is a takeoff from a primary system; it does not circulate directly through the source of heat or cooling but only indirectly through a heat exchanger.

Secondary Treatment. Treatment of boiler feedwater or internal treatment of boiler water after primary treatment.

Sedimentation. Gravitational settling of solid particles in a liquid system.

Seebeck Effect. When two different adjacent metals are heated, an electric current is generated between the metals.

Selective Catalytic Reduction (SCR). Term frequently used as a synonym for catalytic reduction of NO_x in diesel exhaust or flue gases by nitrogen containing compounds, such as ammonia or urea. Such SCR systems are commercially available for stationary applications and are being developed for mobile diesel engines. Since "selective catalytic reduction" is a generic term which also applies to other reactions, its use may lead to confusion in some situations.

Sensible Heat Factor. The ratio of sensible heat to total heat.

Sensible Heat. 1. Heat which causes a change in temperature of a substance. 2. Sensible heat is any heat transfer that causes a change in temperature. Heating and cooling of air and water that may be measured with a thermometer is sensible heat. Heating or cooling coils that simply increase or decrease the air temperature without a change in moisture content are examples of sensible heat.

Sensing Device. A device that keeps track of the measured condition and its fluctuations so that when sufficient variation occurs it will originate the signal to revise the operation of the system and offset the change. Example: a thermostat "bulb". A sensing de vice may be an integral part of a controller.

Sensing Element. The first system element or group of elements. The sensing element performs the initial measurement operation.

Separator. A tank type pressure vessel installed in a steam pipe to collect condensate to be trapped off and thus providing comparatively dry steam to the connected machinery.

Separator. Device to separate one substance from another.

Sequencer. A mechanical or electrical device that may be set to initiate a series of events and to make the events follow in sequence.

Sequester. To form a stable, water-soluble complex.

Sequestrant. A substance that holds a mineral or metal in solution beyond its saturation point.

Series Circuit. One with all the elements connected end to end. The current is the same throughout but the voltage can be different across each element.

Service Drop. The overhead service wires that serve a building.

Service Switch. Disconnect switches or circuit breakers. Purpose is to completely disconnect the building from the electric service.

Service Valve. Manually operated valve mounted on refrigerating systems used for service operation.

Set Point. The value of the controlled condition at which the instrument is set to operate.

Settling Basin. A containment design with external water treatment to settle sediments and to clarify.

Shaft Seal. Device used to prevent leakage between shaft and housing.

Sharp Freezing. Refrigeration at temperatures slightly below freezing, with moderate air circulation.

Shed Mode. A method of demand control that reduces kW demand through shedding and restoring loads.

Shed. To de-energize a load in order to maintain a kW demand set point.

Shell and Tube Flooded Evaporator. Device which flows water through tubes built into cylindrical evaporator or vice versa.

Shell Type Condenser. Cylinder or receiver which contains condensing water coils or tubes.

Shell-And-Tube. Designation of a heat exchanger having straight tubes encased inside a shell.

Shielded Cable. Special cable used with equipment that generates a low voltage output. Used to minimize the effects of frequency "noise" on the output signal.

Shock Dosage. The feeding of treatment to a system all in one slug or dose rather than gradually (also called a slug dosage).

Shock Feeder. A device which is used to add treatment to a system in an instantaneous manner.

Short Circuit. A direct connection of low resistive value that can significant alter the behavior of an element or system.

Short Cycling. Unit runs and then stops at short intervals; generally this excessive cycling rate is hard on the system equipment.

Shroud, Pumps. The front and/or back of an impeller.

Shroud, Turbines. Also referred to as the sealing strip on turbine blades. Its purpose is to supply rigidity to the blades, lessen vibration and provide sealing between stages.

Shunt. A device to divert part of an electric current.

Sight Glass. Glass tube or glass window in refrigerating mechanism. It shows amount of refrigerant or oil in system and indicates presence of gas bubbles in liquid line.

Silica Gel. Absorbent chemical compound used as a drier. When heated, moisture is released and compound may be reused.

Silica, Colloidal. Silica in colloidal form.

Silica, Volatile. Silica carryover with steam.

Silica. Silicon dioxide ($Si0_2$), a mineral found naturally as quartz or in complex combination with other elements such as silicates.

Silt Density Index. A measure of the tendency of a water to foul a reverse osmosis membrane, based on time flow through a membrane filter at constant pressure.

Simple Cycle. Referring to the gas turbine cycle consisting only of compression, combustion and expansion.

Single Phasing. The condition when one phase of a multiphase (poly-phase) motor circuit is broken or opened. Motors running when this occurs may continue to run but with lower power output and over heating.

Single Shaft Gas Turbine. A gas turbine arrangement in which the compressor and the gas turbine are all coupled to one shaft.

Single Stage Compressor. Compressor having only one compressive step between inlet and outlet.

Sintering. Densification of a particulate ceramic compact involving a removal of the pores between the starting particles (accompanied by equivalent shrinkage) combined with coalescence and strong bonding between adjacent particles.

Size Selective Sampling (SSS). A method of measuring ambient DPM exposures through a mechanical separation (inertial impaction) of particles into two fractions: above and below (typically) 0.8 µm, followed by gravimetric analysis. Based on air quality research in underground mining, it is believed that the particle fraction below the 0.8 µm cut point can be attributed to diesel exhaust.

Skin Condenser. Condenser using the outer surface of the cabinet as the heat radiating medium.

Slime. A soft, sticky, mucus-like substance, originating from a bacterial growth.

Sling Psychrometer. Measuring device with wet and dry bulb thermometers. Moved rapidly through air it measures humidity.

Sludge. A deposit on a heat-transfer surface that does not have the hard, crystalline structure of a scale but is softer and less dense.

Slug. A unit of measure for mass in the English system, which equals 14.6 kg in the SI system.

Slugging. Condition in which mass of liquid enters compressor causing hammering.

Slurry Erosion. Material removal due to the combined action of corrosion and wear.

Slurry. A water containing high concentration of suspended solids, usually over 5000 mg/L.

Soda Ash. A common water treatment chemical, sodium carbonate.

Sodium Chloride. Common table salt, used to produce a brine solution, used a secondary refrigerant.

Sodium Sulfite (Na_2sO_3). A chemical used with water treatment to remove small amounts of oxygen.

Sodium Tracer Method. A technique used to measure dissolved solids in steam to values as low as 0.001 ppm.

Sodium Zeolite Softening. The process of removing scale forming ions of calcium and magnesium and replacing them with the equivalent amount of sodium ions.

Soft Water. Water that is free of magnesium or calcium salts.

Softening. The removal of hardness (calcium and magnesium) from water.

Solar Heat. Heat created by visible and invisible energy waves from the sun.

Solder Balls. Small spheres of solder adhering to the laminate, mask, or conductor surfaces usually after wave or reflow soldering.

Solder Bridging. When solder paste or solder on two or more adjacent pads come into contact to form a conductive path or bridge.

Solder Mask. A dielectric material used to cover the entire surface (except where the joints are to be formed) of a PCB primarily to protect the circuitry from environmental damage; it also helps to reduce bridging.

Solder Paste. Mixture of minute spherical solder particles, activators, solvent, and a gelling or suspension agent.

Solder. A low melting point alloy, usually of lead (Pb) and tin (Sn), that can wet copper, conduct current, and mechanically join conductors.

Soldering. Process of joining metallic surfaces with solder without melting the base material.

Solenoid Valve. Electromagnet with a moving core. It serves as a valve or operates a valve.

Solid Absorbent Refrigeration. Refrigeration system which uses solid substance as absorber of the refrigerant during the cooling part of the cycle and releases refrigerant when heated during generating part of cycle.

Solid Solution. A single, solid, homogeneous crystalline phase containing two or more chemical species.

Solid State Halogen Leak Detector. An electronic leak detector for all halogen. related refrigerants.

Solidification Range. The temperature between the liquidus and solidus.

Solidification Shrinkage Crack. A crack that forms due to internal stresses developed from shrinkage during solidification of a metal casting; also called a hot crack.

Solidification. The change from liquid state to solid state upon cooling through the melting temperature or melting range.

Solidus. The temperature in a phase equilibrium diagram below which no liquids are present; the highest temperature at which a metal or alloy is completely solid.

Soluble Iron. Usually present in cooling water systems and can arise from metallurgical corrosion.

Soluble Organic Fraction (SOF). The organic fraction of diesel particulates. SOF includes heavy hydrocarbons derived from the fuel and from the engine lubricating oil. The term "soluble" originates from the analytical method used to measure SOF which is based on extraction of particulate matter samples using organic solvents.

Solu-Bridge. An electronic instrument used to measure conductivity of a water sample to determine the dissolved solids content.

Solution Heat Treatment. Heating an alloy to a suitable temperature, holding at that temperature for enough time to allow one or more constituents to enter into solid solution, and then rapidly cooling to hold the consituents in solution.

Solvent. Solution capable of dissolving a solute.

Solvus. The curve on a phase equilibrium diagram that defines the limits of solid solubility.

Sorbent. See absorbent.

Sour Gas. A gaseous environment containing hydrogen sulfide and carbon dioxide in hydrocarbon reservoirs.

Sour Water. Waste water containing malodorous materials, usually sulfur compounds.

Sparger. An extension into the bottom of a tank of a pipe which has a distribution nozzle on the end for mixing one fluid with another.

Specific Conductance. Measures the ability of a water to conduct electricity. Conductivity increases with total dissolved solids and is therefore used to estimate dissolved solids present in the water.

Specific Gravity. The density of a substance compared to the density of a standard material such as water.

Specific Heat (Cp). The ratio of the mount of heat required to raise a mass of material 1 degree in temperature to the amount required to raise n eual mass of reference substance, usually water, 1 degree in temperature.

Specific Heat. Ratio of quantity of heat required to raise temperature of a body 1 degree to that required to raise temperature of equal mass of water 1 degree.

Specific Humidity. Ratio of weight of vapor to the weight of gas in a unit volume of an air-water vapor mixture.

Specific Volume. Volume per unit mass of a substance.

Spinodal Structure. A fine homogeneous mixture of two phases that form by growth of composition waves in a solid solution during suitable heat treatment; the phases differ in composition from each other and the parent phase, but have the same crystal structure as the parent phase.

Splash System, Oiling. Method of lubricating moving parts by agitating or splashing oil in the crankcase.

Split-Stream Dealkalizer. Where the flow of water is divided through a parallel arrangement of hydrogen and sodium cation exchanger. The combined product being soft and low in alkalinity.

Spray Carryover. Are referred to a mist or fog and are a degree of atomization of the boiler water and carried with the steam. This type of carryover is to be prevented by the drum internals.

Spray Cooling. Method of refrigerating by spraying expendable refrigerant or by spraying refrigerated water.

Spray Manifold. A pipe, or extension of a water line, that has several openings fitted with nozzles which spray water.

Spray-Coil Unit. A cooling circuit that sprays water over cooling coils through inflowing air to humidify or dehumidify that air, as required.

Spray-Cooling Circuit. An open cooling-water circuit which sprays water and cools by evaporation, for example, a cooling tower, all evaporative condenser, an air washer, or a spray-coil unit.

Spread. The divergence of the air stream in a horizontal or vertical plane after it leaves the outlet.

Stability Index. An imperical modification of the saturation index used to predict scaling or corrosive tendencies in water systems.

Stainless Steel. Any steel containing at least 10.5% Cr as the principal alloying element.

Standard Air Conditions. Standard air density has been set at 0.075 Ib/cu ft. This corresponds approximately to dry air at 70□F and 29.92 in Hg. In metric units, the standard air density is 1.2041 kg/m^3 at 20□C and at 101.325 kPa.

Standard Conditions. The standard conditions referred to in environmental system work for air are: dry air at 70□F and at an atmospheric pressure of 29.92 inches mercury (in Hg). For water, standard conditions are 68□F at the same barometric pressure. At these standard conditions, the density of air is 0.075 pounds per cubic feet and the density of water is 62.4 pounds per cubic foot.

State. Refers to the form of a fluid, either liquid, gas or solid. Liquids used in environmental systems are water, thermal fluids such as ethylene glycol solutions, and refrigerants in the liquid state. Gases are steam, evaporated refrigerants and the air-water vapor mixture found in the atmosphere. Some substances, including commonly used refrigerants, may exist in any of three states. A simple example is water, which may be solid (ice), liquid (water), or gas (steam or water vapor).

Static Head. The pressure due to the weight of a fluid above the point of measurement.

Static Suction Head. The positive vertical height in feet from the pump centerline to the top of the level of the liquid source.

Static Suction Lift. The distance in feet between the pump centerline and the source of liquid below the pump centerline.

Steam Drum. A pressure chamber located at the upper extremity of a boiler circulatory system, in which the steam is generated in the boiler and separated from the water.

Steam Jet Refrigeration. Refrigerating system which uses a steam venturi to create high vacuum (low pressure) on a water container causing water to evaporate at low temperature.

Steam Purity. Refers to all matter but water in the steam.

Steam Quality. The percentage by weight of vapor in a steam and water mixture.

Steam. Water in vapor state.

Steam-Absorption Condenser. That part of a steam-absorption machine in which the water refrigerant is condensed by cooling-tower water and returned to the evaporator or chiller.

Steam-Absorption Machine. A refrigeration or air-conditioning machine which uses, as a refrigerant, water evaporated by absorption in a brine regenerated by steam and condensed by cooling-tower water.

Steel. An iron-based alloy containing manganese, usually carbon, and other alloying elements.

Stiction (Static Friction). Resistance of start of motion.

Stoichiometric. The ratio of chemical substances reacting in the water that correspond to their combining weights in the theoretical chemical reaction.

Stoichiometry. Refers generally to the compositior of a material and specifically to the relative atomic proportions of cations and anions.

Strain. The unit of change in the size or shape of a body due to force; a dimensionless number that characterizes the change in dimensions of an object during a deformation or flow process.

Stratified Air. Unmixed air in a duct that is in thermal layers that have temperature variations of more than five degrees.

Stress Intensity Factor (*K*). A scale factor to define the magnitude of the crack-tip stress field; the functionality depends on the configuration of the cracked component and the manner in which the loads are applied. (see Griffith crack theory)

Stress Rapture. A general type of damage referring to carbon steel tubing, when heated above 450□C. Material will plastically deform (creep) and then rapture.

Stress. Force per unit area.

Stuffing Box. That portion of the pump which houses the packing or mechanical seal, The stuffing box is usually referred to as the dry portion of the pump, and is located in back of the impeller and around the shaft.

Subcooling. The difference between the temperature of a pure condensable fluid below saturation and the temperature at the liquid saturated state, at the same pressure.

Subcooling. The process of cooling a liquid to a temperature below its saturation temperature for any given saturation pressure.

Sublimation. A change of state directly from solid to gas without appearance of liquid.

Sublimation. Condition where a substance changes from a solid to a gas without becoming a liquid.

Suction Head. The positive pressure on the pump inlet when the source of liquid supply is above the pump centerline.

Suction Lift. 1. A pumping condition where the eye of the impeller of the pump is above the surface of the water from which the pump is pumping.

2. The combination of static suction lift and friction head in the suction piping when the source of liquid is below the pump centerline.

Suction Line. Tube or pipe used to carry refrigerant gas from evaporator to compressor.

Suction Pressure Control Valve. 1. Device located in the suction line which maintains constant pressure in evaporator during running portion of cycle.

2. Device located in the suction line which maintains constant pressure in evaporator during running portion of cycle.

Suction Pressure. Pressure in low-pressure side of a refrigerating system.

Suction Service Valve. Two-way manually operated valve located at the inlet to compressor. It controls suction gas flow and is used to service unit.

Suction Side. Low-pressure side of the system extending from the refrigerant control through the evaporator to the inlet valve of the compressor.

Sulfate. A compound, ion, or salt of sulfur and oxygen, such as sodium sulfate (Na_2S0_4).

Sulfite Decomposition. Sodium sulfite, which is used as an oxygen scavenger, may decomposes with higher temperatures and concentration. The decomposition results in forming sulfur dioxide and thus leading to an acidic anhydride causing corrosion.

Sulfonic. A specific acidic group (SO_3H) on which depends the exchange activity of certain cation adsorbents.

Sulfur Dioxide (SO_2). An old refrigerant.

Sun Effect. Solar energy transmitted into space through windows and building materials.

Super Ultra Low Emission Vehicle (SULEV). A vehicle that meets the SULEV emission standards set by the California Air Resources Board. The SULEV standards are more stringent than the ULEV standards.

Supercharger. A compressor device to compress the combusti air or the air/fuel mixture before it enters the engine cylind Superchargers are typically driven by the engine itse through a system of gears, a belt drive, or by an electri motor.

Superheat. The heat added to a fluid above its saturation poi

Superheated Steam. Steam heated above its saturati temperature.

Superheated Vapor. A vapor which is not about to condense.

Superheater. Heat exchanger arranged to take heat from liqu going to evaporator and using it to superheat vapor leavi evaporator.

Superheating. The process of adding heat to a vapor in order raise its temperature above saturation temperature. It impossible to superheat a saturated vapor as long as it is contact with the liquid from which it is being generated; hen the vapor must be led away from the liquid before it can superheated.

Supersaturated Solution. A solution of a salt or mineral with concentration beyond the normal saturation point.

Surface Blowdown. Removal of water, foam, etc. from the surfa at the water level in a boiler.

Surface Heating. The exterior surface of a heating unit. Extende heating surface (or extended surface), consisting of fins, pin or ribs which receive heat by conduction from the prim surface. Prime surface: heating surface having the heatir medium on one side and air (or extended surface) on th other.

Surface Insulation Resistance. The electrical resistance of a insulating material between a pair of contacts or conductor SIR is determined under specified environmental an electrical conditions.

rface Mount Technology. Method of assembling printed circuit boards where the components are mounted onto the surface of the board rather than being inserted into holes in the board.

rface-Spray Unit. A spray-coil unit.

rfactant. A compound that affects interfacial tension between two liquids. It usually reduces surface tension.

rfactants. A wetting agent used to prevent fouling, mainly in water cooling systems.

ge Surpressor. A device that reduces harmonic distortion in line voltage circuits by clipping off transient voltages which are fed through the power lines from operating equipment.

ge Tank. Container connected to the low-pressure side of a refrigerating system which increases gas volume and reduces rate of pressure change.

ge. The sudden displacement or movement of water in a closed vessel or drum.

pended Solids. Un-dissolved solids in boiler water.

amp Cooler. Evaporative type cooler in which air is drawn through porous mats soaked with water.

lling. The expansion of an ion-exchange which occurs when the reactive groups on the resin are converted from one form to another.

rl Combustion. A combustion chamber configuration which uses curved mixing ridges in the intake ports and/or a re-entrant piston bowl (a bowl whose top edges curve inward). Some swirl combustion chambers have a larger rim around the outside of the piston and a more compact combustion chamber or bowl. The swirl is used to reduce particulate emissions.

ergism. The combined action of several chemicals which produce an effect greater than the additive effects of each.

System, Central Fan. A mechanical, indirect system of heating ventilating, or air conditioning, in which the air is treated o handled by equipment located outside the rooms served usually at a central location, and conveyed to and from th rooms by means of a fan and a system of distributing ducts

System, Closed. A heating or refrigerating piping system in which circulating water or brine is completely enclosed, unde pressure above atmospheric, and shut off from th atmosphere except for an expansion tank.

System, Duct. A series of ducts, conduits, elbows, branch piping etc. designed to guide the flow of air, gas or vapor to an from one or more locations. A fan provides the necessar energy to overcome the resistance to flow of the system an causes air or gas to flow through the system. Som components of a typical system are louvers, grilles, diffusers filters, heating and cooling coils energy recovery de vices burner assemblies, volume dampers, mixing boxes, soun attenuators, the ductwork and related fittings.

System, Flooded. A system in which only part of the refrigeran passing over the heat transfer surface is evaporated, and th portion not evaporated is separated from the vapor an recirculated.

System, Unitary. A complete, factory-assembled and factory-teste refrigerating system comprising one or more assemblie which may be shipped as one unit or separately but whic are designed to be used together.

System. A series of ducts, conduits, elbows, branch piping, et designed to guide the flow of air, gas or vapor to and fror one or more locations. A fan provides the necessary energ to overcome the resistance to flow of the system and cause air or gas flow through the system. Some components of typical system are louvers, grilles, diffusers, filters, heatin and cooling coils, air pollution control devices, burne assemblies, volume flow control dampers, mixing boxe sound attenuators, the ductwork and related fittings.

Systems Curve. A graphic presentation of the pressure vs. volum flow rate characteristics of a particular system.

T

Tall PIPE. Outlet pipe from the evaporator.

Tandem Compound Turbine. turbines are large turbines consisting of two or more turbines in series coupled together as one shaft and applied to one generator

Tannins. A chemical used as an inhibitor in relation with caustic embrittlement.

Temperature Control. Temperature-operated thermostatic device which automatically opens or closes a circuit.

Temperature Critical. The saturation temperature corresponding to the critical state of the substance at which the properties of the liquid and vapor are identical.

Temperature, Absolute Zero. The zero point on the absolute temperature scale, 459.69 degrees below the zero of the Fahrenheit scale, 273.16 degrees be low the zero of the Celsius scale.

Temperature, Dewpoint. The temperature at which the condensation of water vapor in a space begins for a given state of humidity and pressure as the temperature of the vapor is reduced. The temperature corresponding to saturation (100 percent relative humidity) for a given absolute humidity at constant pressure.

Temperature, Drybulb. The temperature of a gas or mixture of gases indicated by an accurate thermometer after correction for radiation.

Temperature, Effective. An arbitrary index which combines into a single value the effect of temperature, humidity, and air movement on the sensation of warmth or cold felt by the human body. The numerical value is that of the temperature of still, saturated air which would induce an identical sensation.

Temperature, Saturation. The temperature at which no further moisture can be added to the air water vapor mixture. Equals dew point temperature.

Temperature, Wet Bulb. Thermodynamic wet bulb temperature is the temperature at which liquid or solid water, by evaporating into air, can bring the air to saturation adiabatically at the same temperature. Wet bulb temperature (without qualification) is the temperature indicated by a wet bulb psychrometer constructed and used according to specifications.

Temperature. Degree of hotness or coldness as measured by a thermometer.

Temperature-Humidity Index. Actual temperature and humidity of air sample compared to air at standard conditions.

Tensile Strength. In tensile testing, the ratio of maximum load to original cross-sectional area. Also called ultimate strength.

Tensile Stress. A stress that causes two parts of an elastic body, on either side of a typical stress plane, to pull apart.

Tension. The force or load that produces elongation.

Terminal Velocity. The maximum air stream velocity at the end of the throw.

The First Law. (1) When work is expanded in generating heat, the quantity of heat produced is proportional to the work expended; and, conversely, when heat is employed in the performance of work, the quantity of heat which disappears is proportional to the work done (Joule); (2) If a system is caused to change from an initial state to a final state by adiabatic means only, the work done is the same for all adiabatic paths connecting the two states (Zemansky); (3) In any power cycle or refrigeration cycle, the net heat absorbed by the working substance is exactly equal to the net work done.

The Second Law. (1) It is impossible for a self acting machine, unaided by any external agency, to convey heat from a body of lower temperature to one of higher temperature (Clausius); (2) It is impossible to derive mechanical work from heat taken from a body unless there is available a body of lower temperature into which the residue not so

Therm. Measurement used by gas utilities for billin3 purposes. 1 Therm = 100 cubic feet of gas = 100,000 Btu.

Therm. Quantity of heat equal to 100000 Btu.

Thermal Conductivity. The rate at which heat is transferred through an object.

Thermal Conductivity. The rate of heat flow, under steady conditions, through unit area, per unit temperature gradient in the direction perpendicular to the area. It is given in the SI nits s watts per meter Kelvin (W/m • K).

Thermal Efficiency of a Gas Turbine. Is the energy output of the gas turbine divided by the energy input of the gas turbine.

Thermal Efficiency. Ratio of shaft work out of a system to the heat energy into the system.

Thermal Electromotive Force. The electromotive force generated in a circuit containing two dissimilar metals when one junction is at temperature different from that of the other. (See also thermocouple).

Thermal Equilibrium. When two bodies originally at different temperatures, have attained the same temperature

Thermal Expansion Valve. The metering device or flow control which regulates the amount of liquid refrigerant which is allowed to enter the evaporator.

Thermal Expansion. 1. Change in dimensions of a material resulting from a change in temperature. 2. The change in length of a material with change in temperature.

Thermal Shock. 1. Stresses induced in a material because of a rapid temperature change or a thermal gradient.

2. The development of a steep temperature gradient and accompanying high stress within a material or structure.

Thermal Treatment. Refers to the treatment of water with heat to drive off the dissolved gases and soften certain minerals for easy removal.

Thermistor. A two-terminal semiconductor device whose resistance is temperature sensitive.

Thermobank. A bank for storing heat.

Thermocouple Thermometer. Electrical instrument using thermocouple as source of electrical flow, connected to millimeter calibrated in temperature degrees.

Thermocouple. Device for measuring temperature utilizing the fact that an electromotive force is generated whenever two junctions of two dissimilar metals in an electric circuit are at different temperature levels.

Thermocouple. 1. Device which generates electricity, using the principle that if two unlike metals are welded together and junction is heated, voltage will develop across the open ends.

2. Device which generates electricity, using the principle that if two unlike metals are welded together and junction is heated, voltage will develop across the open ends.

Thermodisk Defrost Control. Electrical switch with bimetal disk controlled by temperature changes.

Thermodynamic Properties. Basic qualities used in defining the condition of a substance, such as temperature, pressure, volume, enthalpy, entropy.

Thermodynamics. Part of science which deals with the relationships between heat and mechanical action.

Thermoelectric Refrigeration. Where refrigeration is produced by the passage of electric current through two dissimilar materials.

Thermoelectricity. In physics, electricity generated by the application of heat to the junction of two dissimilar materials. If two wires of different materials are joined at their ends and one end is maintained at a higher temperature than the other, a voltage difference will arise, and an electric current will exist between the hot and the cold junctions.

Thermometer. Device for measuring temperatures.

Thermomodule. Number of thermocouples used in parallel to achieve low temperatures.

Thermopile. Number of thermocouples used in series to create a higher voltage.

Thermostat. Device, which senses ambient temperature, conditions and, in turn, acts to control a circuit.

Thermostatic Control. Device which operates system or part of system based on temperature change.

Thermostatic Valve. Valve controlled by temperature change response elements.

Thermostatic Water Valve. Valve used to control flow of water through system, actuated (made to work) by temperature difference. Used in units such as water-cooled compressor and/or condenser.

Three-Way Valve. Multi-orifice (opening) flow control valve with three fluid flow openings.

Threshold Limit Value (TLV). Time-weighted average concentration of an air pollutant at the workplace for a conventional 8-hour workday and a 40-hour workweek, to which nearly all workers may be repeatedly exposed without adverse health effects.

Threshold Treatment. Chemical treatment, used to prevent scale formation, which acts to hold hardness in solution at the threshold of precipitation.

Throttle Governing. With throttle governing a single large control valve controls the load from 0% to 100%When steam is throttled, the superheat increases and the turbine exhaust steam is drier, reducing the turbine blade erosion, but with the drier steam entering the condenser, the condenser losses increase. Throttling of steam through a valve is an isenthalpic (constant enthalpy) process and no heat is lost. The so-called throttling losses occur in the condenser.

Throttling Range. The amount of change in the variable being controlled to make the controlled device more through the full length of its stroke.

Throttling. An irreversible adiabatic steady flow process in which the fluid is caused to flow through an obstruction in a pipe with a resulting drop in pressure.

Thrust Collar Position Indicator. The axial position of the rotor is very important and an axial position indicator is often applied to the thrust bearing. It may be a large dial micrometer with alarm setting for an axial movement of 0.4 millimeter and shutdown at 0.8 millimeter, or An oil pressure gauge connected to an oil leak-off device may also be used as an axial position indicator. The oil is supplied at say 500 kPa, flows through an orifice and leaks off through a nozzle. The pressure between the orifice and nozzle depends on the distance between the nozzle and shaft thrust collar; the larger the distance the lower the pressure. The pressure gauge can be calibrated in millimeter clearance and may have alarm and shutdown settings

Timer-Thermostat. Thermostat control which includes a clock mechanism. Unit automatically controls room temperature and changes temperature range depending on time of day.

Titration. A chemical process used in analyzing feed water.

Ton Refrigeration Unit. Unit which removes same amount of heat in 24 hours as melting of 1 ton of ice.

Ton's of Refrigeration. The capacity of a refrigeration system that can freeze 1 ton (1000 kg) of liquid water at 0□C into ice at 0□C in 24 hour is said to be 1 tone.

Tool Steel. Any steel used o make tools for cutting, forming, or otherwise shaping a material into a final part.

Topping Turbine. Have been used when old boilers are replaced with new high pressure boilers. The turbine is a backpressure turbine exhausting to the old boiler header still supplying steam to the old lower pressure turbines.

Total Carbon (TC). The sum of the elemental carbon and organic carbon associated with diesel particulates. Typically amounts to 80-85% of the total DPM mass.

Total Dynamic Head. Dynamic discharge head (static discharge head, plus friction head, plus velocity head) plus dynamic suction lift, or dynamic discharge head minus dynamic suction head.

Total Hardness. See Hardness.

Total Heat (Enthalpy). Total heat is the sum of the sensible heat and latent heat in an exchange process. In many cases, the addition or subtraction of latent and sensible heat at terminal coils appears simultaneously. Total heat also is called enthalpy, both of which can be defined as the quantity of heat energy contained in that substance.

Total Heat. Sum of both the sensible and latent heat.

Total Particulate Matter (TPM). The total particulate matter emissions including all fractions of diesel particulates, i.e. the carbonaceous, organic (SOF), and sulfate particulates.

Total Solids. Are the sum of the dissolved and suspended solids.

Toughness. A measure of the energy absorbed before and during the fracture process; it is equal to the area under the tensile stress-strain curve.

Tower Fill. The interior structure of a cooling tower over which the water flows.

Trace Constituents. Materials present at a concentration less than 0.01 mg/L.

Transducer. The means by which the controller converts the signal from the sensing device into the means necessary to have the appropriate effect on the controlled device. For example, a change in air pressure in the pneumatic transmission piping.

Transformer. The system power supplying transformer is an inductive stationary device which transfers electrical energy from one circuit to another. The transformer has two windings, primary and secondary. A changing voltage applied to one of these, usually the primary, induces a current to flow in the other winding. A coupling transformer transfers energy at the same voltage; a step-down transformer transfers energy at a lower voltage, and a step-up transformer transfers energy at a higher voltage.

Transgranular. Through or across crystals or grains; also called intracrystalline or transcrystalline.

Transient Constituents. Are those constituents which change in concentration or activity by changes in the aquatic environment. The change my be due to oxidation potentials, biological activities, etc.

Transistor. An active semiconductor device capable of providing power amplification and having three or more terminals.

Transition Metal. A metal having available electron energy levels occupied so that the *d*-band contains less than its maximum number of ten electrons per atom; the incompletely filled *d*-levels cause the unique properties of the transition metals; included in this class are iron, cobalt, nickel, and tungsten.

Transition Temperature. An arbitrarily defined temperature that lies within the temperature range where metal fracture characteristics change rapidly; an example is the ductile-to-brittle transition temperature (DBTT).

Transitional Low Emission Vehicle (TLEV). A vehicle that is certified to meet the TLEV emission standards established by the California Air Resources Board (CARB). The TLEV standards are less stringent than the LEV standards.

Transmittance, Thermal (U Factor). The time rate of heat flow per unit area under steady conditions from the fluid on the warm side of a barrier to the fluid on the cold side, per unit temperature difference between the two fluids.

Treatment. A process whereby impurities are removed from water; also a substance added to water to improve its physical or chemical properties.

Tribology. 1. The science and technology of interacting surfaces in relative motion and of the practices related thereto; the science concerned with the design, friction, lubrication, and wear of contacting surfaces that move relative to one another (e.g. Bearings, cams, gears).

2. The science concerned with the design, friction, lubrication and wear of contacting surfaces that move relative to each other (as in bearings, cams, or gears).

Triclinic. Having three axes of any length with none of the included angles being equal to one another or 90 degrees.

True Strain. The ratio of the change in dimension, resulting from a given load, to the magnitude of the dimension immediately prior to applying the load; the natural logarithm of the ratio of the gage length at the moment of observation to the original gage length.

True Stress. The load applied to a material divided by the cross sectional area over which it acts.

Tube Sheet. The portion of a heat exchanger or boiler in to which the tubes are rolled or secured.

Tubercle. A protective crust of corrosion products (rust) which builds up over a pit caused by the loss of metal due to corrosion.

Tuberculation. A corrosion process that produces hard knob-like mounds of corrosive products on metal surfaces, increasing friction and reducing flow in a water distribution system.

Tube-Within-A-Tube. Water-cooled condensing unit in which a small tube is placed inside large unit. Refrigerant passes through outer tube, water through the inner tube.

Turbidity Unit. The unit of measure of suspended matter in water. It is the measure of light compared against light reflected by a reference standard as defined by the standard methods of water analysis in, APHA.

Turbidity. The measure of suspended matter in, a water sample which contributes to the reflection of light or cloudiness.

Turbine Meter. A device used to measure water consumption in industrial plants.

Turbine Rotor. The rotating assembly enclosed within the turbine casing

Turbine. A rotary machine which extracts mechanical shaft power from the working fluid (gas or liquid) using rotor vanes.

Turbo Generator. Gas turbine combined with an electrical generator.

Turbo Lag. The time delay between injecting fuel to accelerate and delivering air to the intake manifold by the turbocharger. This phenomenon may cause black smoke emissions in some turbocharged diesel engines during acceleration.

Turbocharging. A process of compressing the engine intake air charge in order to allow more air and fuel into the cylinder and, thus, to increase the engine power output. The compressor, called the turbocharger, is driven by an exhaust gas propelled turbine.

Turner Gauge. A device used to measure the actual scale-thickness in boiler tubes.

Twin Bands. Bands across a crystal grain where crystallographic orientations have a mirror image relationship to the orientation of the matrix grain across a composition plane usually parallel to the sides of the band.

Twin. Two portions of a crystal with a definite orientation relationship; the orientation of one portion is a mirror image of the orientation of the other portion across a twinning plane or an orientation derived by rotating one portion about a twinning axis.

Two-Temperature Valve. Pressure-opened valve used in suction line on multiple refrigerator installations, which maintains evaporators in a system at different temperatures.

Two-Way Valve. Valve with one inlet port and one outlet port.

U

Ultimate Strength. The maximum stress (tensile, compressive or shear) a material can sustain without fracture. It is determined by dividing maximum load by the original cross-sectional area of the specimen.

Ultra Filtration. A process that forces water through a filtering membrane by means of pressure gradients in order to obtain ultra pure water.

Ultra-Low Emission Vehicle (ULEV). A vehicle that meets the ULEV emission standards set by the California Air Resources Board. The ULEV standards are more stringent than the LEV standards.

Under Deposit Attack. Corrosion under or around a localized deposit on a metal surface (a form of crevice corrosion).

Under-cooling. Cooling a material below the temperature of an equilibrium phase change fast enough to not allow the occurrence of the transformation.

Unit Cell. The smallest repetitive volume that comprises the complete pattern of a crystal.

Unit Injector. An injector which is camshaft-driven and incorporates a plunger. The plunger works in conjunction with orifices in the injector body to determine the beginning and end of injection. The plungers has a helix and is rotated by a rack so the beginning and end of injection can occur closer together or farther apart, thus shortening or lengthening the injection period and changing metering. A special pump which supplies fuel through an orifice to the injector operates at a pressure which is precisely controlled depending on the changes in engine speed.

Unitary System. A room unit which performs part or all of the air conditioning functions. It may or may not be used with a central fan system.

Unloader. A device in or on the compressor for equalizing high-side and low-side pressures for a brief time during starting and for controlling compressor capacity by rendering one or more cylinders ineffective.

Upflow Filter. A unit containing a single filter medium, usually with graded sand.

Upflow Furnace. A furnace in which the heated air flows upward as it leaves the furnace.

Upflow. The operation of an ion-exchange unit in which solutions are passed in at the bottom and out at the top of the container.

Upstream. The inlet side of an instrument, a pump, valve, etc..

Utility Transformer. Primary and secondary coils of wire which reduce (step down) the utility supply volt age for use within a facility.

U-Tube Manometer. A U-shaped section of plastic or glass tubing that is partially filled with water or mercury. They are used to measure the lower pressure ranges of gases.

V

Vacancy. An unfilled lattice site in a crystal structure.

Vacuum Breaker. A device to prevent a suction in a water pipe.

Vacuum Pump. Special high efficiency device used for creating high vacuums for testing or drying purposes.

Vacuum. Pressure lower than atmospheric pressure.

Valence. The charge on an ion based on the number of electrons transferred or shared within a specific structure.

Valve, Modulating. A valve which can be positioned anywhere between fully on and fully off to proportion the rate of flow in response to a modulating controller (see modulating control).

Valve, Needle. A form of globe valve that contains a sharp pointed, needle like plug that is driven into the and out of a cone shaped seat to accurately control a relatively small rate of flow of a fluid.

Valve, Pop. A spring loaded safety valve that opens automatically when pressure exceeds the limits for which the valve is set. It is used a safety device on pressurized vessels and other equipment to prevent damage from excessive pressure, also called relief valve or a safety valve.

Valve, Poppet. A device that controls the rate of flow of fluid in a line or opens or shuts of the flow of fluid completely. When open, the sealing surface of the valve is moved away from a seat. When closed, the sealing surface contacts the seat to shut of the flow. Poppet valves are used extensively as pneumatic controls and as intake and exhaust valves in most internal combustion engines.

Valve, Pressure Relief. A valve designed to minimize the possibility of explosion when air temperature surrounding a refrigeration system may rise to a point where the pressure of the refrigerant gas to increase to a danger point.

Valve, Relief. Also called pressure relief valve.

Valve, Two-Position. A valve which is either fully on or fully off with no positions between. Also called an "on-off valve".

Vane. That portion of an impeller which throws the water toward the volute case.

Vapor Barrier. A moisture-impervious layer applied to the surfaces enclosing a humid space to prevent moisture travel to a point where it may condense due to lower temperature.

Vapor Lock. A condition where liquid flow is impeded by vapor trapped in a liquid line.

Vapor Phase Inhibitors. A system using an organic nitrite compound, a powder which vaporizes slowly to protect ferrous metal from contact with oxygen.

Vapor Pressure. Vapor pressure denotes the lowest absolute pressure that a given liquid at a given temperature will remain liquid before evaporating into its gaseous form or state.

Vapor, Saturated. Vapor in equilibrium with its liquid; i.e., when the numbers per unit time of molecules passing in two directions through the surface dividing the two phases are equal.

Vapor, Superheated. Vapor at a temperature which is higher than the saturation temperature (i.e., boiling point) at the existing pressure.

Vapor, Water. Water used commonly in air conditioning parlance to refer to steam in the atmosphere.

Vapor. A gas, particularly one near to equilibrium with the liquid phase of the substance and which does not follow the gas laws. Usually used instead of gas for a refrigerant, and, in general, for any gas below the critical temperature.

Vaporous Carryover. Referring to impurities carried over with the steam and then forming a deposit on turbine bladings. This type of carryover is difficult to prevent.

Velocity Head. The pressure needed to accelerate the fluid being pumped.

Velocity Head. The vertical distance a liquid must fall to acquire the velocity with which it flows through the piping system. For a given quantity of flow, the velocity head will vary indirectly as the pipe diameter varies.

Velocity, Terminal. The highest sustained air stream velocity existing in the mixed air path at the end of the throw.

Velocity. A vector quantity which denotes, at once, the time rate and the direction of a linear motion.

Vent. An opening in a vessel or other enclosed space for the removal of gas or vapor.

Ventilation. The process of supplying or removing air by natural or mechanical means, to or from a space; such air may or may not have been conditioned.

Venturi Tube Meter. A flow meter used to determine the rate of flow and employing a venturi tube as the primary element for creating differential pressure in flowing gases or liquids.

Virgin. A metal made directly from ore by smelting.

Viscosity Index. A commonly used measure of the change in viscosity of a fluid with temperature. The higher the viscosity index, the smaller the relative change in viscosity with temperature.

Viscosity. 1. Coefficient of resistance to flow. 2. That property of semi-fluids, fluids, and gases by virtue of which they resist an instantaneous change of shape or arrangement of parts. It is the cause of fluid friction whenever adjacent layers of fluid move with relation to each other.

Vital Heat. The heat generated by fruits and vegetables in storage; caused by ripening.

VMT. Vehicle miles traveled.

Volatile Organic Compounds (VOC). Hydrocarbon-based emissions released through evaporation or combustion. The term VOC is usually used in regard to stationary emission sources.

Volatile Organic Fraction (VOF). The organic fraction of diesel particulate matter as determined by vacuum evaporation. It may or may not be equivalent to the SOF fraction. Depending on the exact analytical procedure, the VOF may include the organic material (SOF) as well as some of the sulfate particulates which, being composed primarily of hydrated sulfuric acid, are also volatile.

Volatile Solids. Those solids in water or other liquids that are lost on ignition of dry solids at 550□F.

Volatile Treatment. Based on the use of hydrazine and neutralizing amines or ammonia. Leaves no solids in the boiler.

Volatility. Volatility, surface tension and capillary action of a fluid are incidental to environmental systems. Volatility is the rapidity with which liquids evaporates extremely rapidly and therefore is highly volatile.

Volt. The unit of potential difference or electromotive force in the meter-kilogram-second system, equal to the potential difference between two points for which 1 coulomb of electricity will do 1 joule of work in going from one point to another.

Voltage (E). The electromotive force in an electrical circuit. The difference in potential between two unlike charges in an electrical circuit is its voltage measured in "volts" (V).

Voltage Drop. The voltage drop around a circuit including wiring and loads must equal the supply volt age.

Voltaic Cell. A storage device that converts chemical to electrical energy.

Volume, Specific. The volume of a substance per unit mass; the reciprocal of density.

Volute. The spiral-shaped casing surrounding a pump impeller that collects the liquid discharged by the impeller.

Walk-In-Cooler. A large commercial refrigerated space often found in supermarkets or places for whole sale distribution.

Waste Water. The used water and solids from industrial processes that flow to a treatment plant.

Water Absorption. The amount of weight gain (%) experienced in a polymer after immersion in water for a specific length of time under controlled environment.

Water Hammer. Banging of pipes caused by the shock of closing valves (faucets).

Water Leg. That space that is full of boiler water between two parallel plates. It usually forms one or more sides of internally fired boilers.

Water Lubricant. Water used as a lubricant; for example, in a mechanical seal on a centrifugal water pump.

Water Softener. A device or system used to remove calcium and magnesium hardness minerals from a water supply.

Water Tube. A boiler tube through which the fluid under pressure flows. The products of combustion surround the tube.

Water Vapor. In air conditioning, the water in the atmosphere.

Water Wall. A row of water tubes lining a furnace or combustion chamber, exposed to the radiant heat of the fire.

Water, Brakish. (1) Water having less salt than sea water, but undrinkable. (2) Water having salinity values ranging from about 0.5 to 17 parts per thousand.

Water, Potable. Water that is safe to drink.

Water, Sour. Waste waters containing fetid materials, usually sulfur compounds.

Water. A tasteless, odorless, colorless liquid in its pure state.

Water-Ice Refrigeration System. Heat is absorbed as ice melts and thus producing a cooling effect.

Watt (W). A measure of electric power equal to a current flow of one ampere under one volt of pressure; or one joule per second in SI units.

Weight to Power Ratio. It is the weight of the machine producing work. For example - the gas turbine is capable of producing more horse power per given mass of its machinery, then the same amount of horse power produced by a machine having many times that mass.

Welding. The joining of two metal surfaces that have been heated, melted, and fused together.

Wet Bulb Temperature (WB). The temperature registered by a thermometer whose bulb is covered by a saturated wick and exposed to a current of rapidly moving air. The wet bulb temperature also represents the dew point temperature of the air, where the moisture of the air condenses on a cold surface.

Wet Bulb. Device used in measurement of relative humidity. Evaporation of moisture lowers temperature of wet bulb compared to dry bulb temperature in same area.

Wet Standby. Boiler is filled completely with water or maintained at normal operating level with a positive nitrogen pressure of 35 to 70 kPa.

Wet-Bulb Depression. The difference between the dry-bulb temperature and the wet bulb temperature.

White Smoke. The smoke emitted during a cold start from a diesel engine, composed mainly of unburnt fuel and particulate matter.

Windage Drift. That water lost from an open re-circulating-water system by means of wind blown through the spray area that carries water out of the system. This is not the same as loss by evaporation, since such a loss can occur even without evaporation.

Wobble Plate-Swash Plate. Type of compressor designed to compress gas, with piston motion parallel to crankshaft.

Work angle. In arc welding, the angle between the electrode and one of the joints.

Work Hardening (strain hardening). Increase dislocation density in metals through straining a material with an applied stress; degree of hardening may be manipulated through recovery and recrystallization.

Working Fluid. is the substance which does the work in a heat engine. The air is one of the working fluids used with gas turbines. Freons are the working fluids used with some refrigeration systems. Water is the working fluid used with steam boilers.

Wrought Alloy. An alloy that is suitable for mechanical forming below melting-point temperatures.

Wrought Iron. Commercial iron that contains less than 0.3% carbon and 1.0 or 2.0% slag, giving it ductility and toughness.

X-radiation. Electromagnetic radiation of the same nature as visible light, but having a wavelength approximately one thousandth that of visible light.

X-ray Tube. A device that produces X-rays by the impact of high-speed electrons on a metal target.

Z

Zeolite Softening. Refers to the process, where zeolite chemicals are capable to exchange ions with the hardness causing impurities of the water.

Zeolite. A natural mineral (hydrous silicates) that has the capacity to absorb hardness, calcium, and magnesium ions from water.

Zero Emission Vehicle (ZEV). A vehicle that is certified to meet the most stringent emission standards established by the California Air Resources Board (CARB). These standards require zero regulated emissions of nonmethane organic gases (NMOGs), carbon monoxide (CO) and nitrogen oxides (NO_x). A ZEV is most likely powered by electricity, fuel cells or hydrogen.

Zeta Potential. The difference in voltage between the surface of the diffuse layer surrounding a colloidal particle and the bulk liquid beyond.

Zone. Any group of crystal planes that are all parallel to one line, called the zone axis.

Zoning. The practice of dividing a building into small sections for heating and cooling control. Each section is selected so that one thermostat can be used to determine its requirements.